D1600660

TAKE A NUMBER

TAKE A NUMBER

Mathematics for the Two Billion

Lillian R. Lieber

Drawings by Hugh Gray Lieber

Dover Publications, Inc.
Mineola, New York

Bibliographical Note

This Dover edition, first published in 2017, is an unabridged republication of the work originally published by The Jaques Cattell Press, Lancaster, Pennsylvania, in 1946.

Library of Congress Cataloging-in-Publication Data

Names: Lieber, Lillian R. (Lillian Rosanoff), 1886–1986. | Lieber, Hugh Gray, 1896– 1961, illustrator.
Title: Take a number : mathematics for the two billion / Lillian R. Lieber ; drawings by Hugh Gray Lieber.
Description: Dover edition. | Mineola, New York : Dover Publications, Inc., 2017. | Originally published: Lancaster, Pennsylvania : The Jaques Cattell Press, 1946.
Identifiers: LCCN 2017005205| ISBN 9780486815589 | ISBN 0486815587
Subjects: LCSH: Mathematics. | Mathematics—Humor.
Classification: LCC QA39 .L489 2017 | DDC 510—dc23 LC record available at https://lccn.loc.gov/2017005205

Manufactured in the United States by LSC Communications
81558701   2017
www.doverpublications.com

TABLE OF CONTENTS

PREFACE

Have you ever been asked to
"Take a Number"
and wondered how
your clever friends could
Read Your Mind?
Well, meet SAM—
your best Guide and Pal—
he will show you how
to solve your difficulties
mathematical et al.

I. WHY SHOULD YOU STUDY
MATHEMATICS?

Most people probably grant nowadays
that
since we live in a
scientific age,
SOMEBODY has to study
Mathematics,
but is this subject necessary for
EVERYONE?
For all the two billion (2×10^9) people
on this earth?
Why not let those who like it
struggle with it,
and let the rest of us
study something else,
or maybe just take it easy
and enjoy life?
We shall try to tell you WHY
everyone, including you,
must study some Mathematics—
and the more the merrier!

And,
with this little book,
perhaps you will even
begin to like it—
we hope!

First of all,
you will admit that
everyone must know
a little Arithmetic,
so that he can at least
count his pennies or his sheep
or what have you,
and carry on
whatever little business
he has to do.
"True," you will say,
"but that is not much.
How about Algebra, though,
and Geometry and all that stuff?
No doubt the scientists need it,
but I am going to be
a businessman, or
a stenographer, or
maybe a housewife,
and I surely don't expect
to use Math.
What I need is
good looks,
money,

common sense—
but not Math."
Now you know perfectly well that
many times
people who are not so good-looking
know how to use
what looks they have
to better advantage than
some very handsome ones.
And that people without money
can go out and make a million
if they have the ability and
if that is what they want.
In short,
what we need most is
a knowledge of
the world we live in and
the "tools" needed to
get along in it.
Now the most important "tool" is
certainly your own brain,
so that
if you learn to use
that "instrument"
you will really have something!
Furthermore,
it is easy to see that
EVERYONE must learn to
use his "bean,"
or else

what one intelligent man or group
builds up
can be destroyed by the
ape-men.
So you see
unless we
ALL GET WISE TOGETHER,
the world cannot be safe for anyone.

Now, since man's brain
has been most successful in
Science (including Physics *et al*),
Art (including Music, Poetry, *et al*), and
Mathematics,
let us see if we can learn from
THEM
how to really use our brains
And, to make it easier,
let us combine them all into
a single character called
SAM—
so that our problem is to help
to bring SAM to the
TWO BILLION,
and show
HOW
he can help us.
You will see that
he can be a better
LEADER

than any individual human being!

Now it must be very clearly understood
that SAM, who is of course
very scientific,
is nevertheless
NOT just a gadgeteer
who furnishes us with
radios and automobiles
and other gadgets
to play with,
though he HAS his pockets full of
toys for us—
but he will give them to us
ONLY if the human race is GOOD—
otherwise he will give us
ATOMIC BOMBS instead!
For in his heart
SAM is really in agreement with
the great RELIGIONS
and will FIGHT for them
and not allow us to
ignore them any longer!

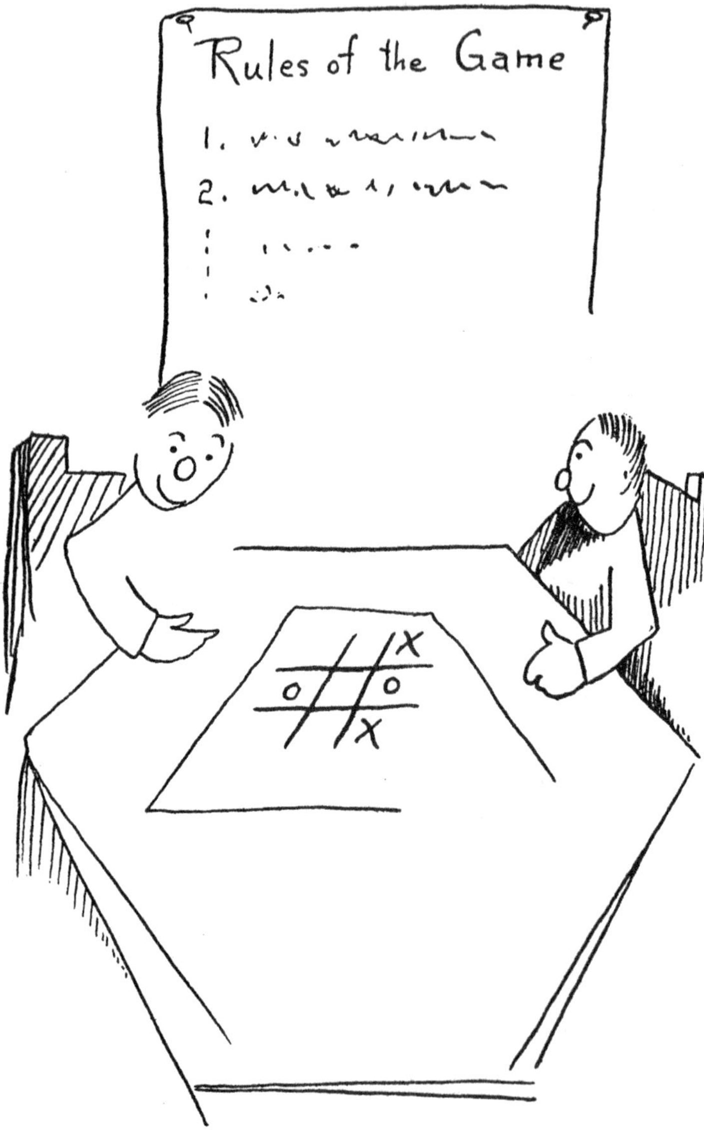
Rules of the Game

II. MATHEMATICS MADE EASY

One of the main difficulties with
the study of Mathematics
by the average person
has always been that
he was never let in on
the basic rules of the game,
but was given
THOUSANDS of little details
to remember,
which did not seem to have
much connection with each other,
so he could never
figure things out for himself.
Can you imagine playing any game
without knowing
the EQUIPMENT of the game or
the RULES?
But just being pushed around
without ever finding out
what you were doing,
or

the difference between
a goal and a foul?!

Now, in this little book,
we shall try to show you that
Mathematics is really like a game.
And if you know
the EQUIPMENT and the RULES
(and you will be surprised to find
how VERY FEW basic rules there are!),
you can easily learn the game
and even figure out your own
"plays."
This not only makes it easy to play,
but you will be surprised to find
what FUN it is!
Try it and see.

III. THE EQUIPMENT

First, take Arithmetic.
The equipment of Arithmetic
consists of
various kinds of NUMBERS,
which we shall recall to your mind
in this chapter.
And you can then easily go from
Arithmetic to Algebra
by merely
ENLARGING this equipment
to include a kind of numbers
called NEGATIVE numbers,*
which are very easy to understand,
as you will see.

* There are really many
  different kinds of Algebras,
  but the one referred to above is
  the one you will read about
  in this book.
  It may be called
  the Algebra of Rational Numbers:
  in this chapter you will soon find out
  just what a
  RATIONAL NUMBER
  is.

Another change in going
from Arithmetic to Algebra
will be in the use of
LETTERS to stand for numbers.
You will see how
this simple device
will make it possible
to solve problems more EASILY
and more GENERALLY.
Although you may not realize it,
you already know, from Arithmetic,
several different kinds of numbers:
for example, as a very small child,
you learned about.
the WHOLE NUMBERS or INTEGERS,
like 1, 7, 11, 79, etc.
Later on,
you also learned about
FRACTIONS,
like $\frac{1}{2}$, $\frac{7}{8}$, $\frac{11}{9}$,
and decimal fractions,
like 1.25, .06, 7.09,
and mixed numbers,
like $1\frac{1}{2}$, $3\frac{1}{4}$,
etc.
Now, fortunately,
it is not necessary to have
so many different names,
for all the numbers mentioned above
are called

RATIONAL NUMBERS.

Let us give you a clear definition of
a RATIONAL NUMBER:
This is
ANY NUMBER WHICH MAY BE EXPRESSED
AS A RATIO* OF TWO INTEGERS:
Thus obviously the fraction $\frac{7}{8}$ is
a rational number since it is
the ratio of
the integer 7 to the integer 8.
Similarly the mixed number $1\frac{1}{2}$ is
a rational number since it can be written $\frac{3}{2}$,
which is the ratio of
the integer 3 to the integer 2.
Also, the decimal .06 is
a rational number, since
it can be written $\frac{6}{100}$ or $\frac{3}{50}$.
And even an integer, like 7,
can be written $\frac{7}{1}$,
which, again, is the ratio of
the integer 7 to the integer 1,
showing that the integers themselves
are also rational numbers.

Now we want to introduce you to
a kind of number which
you have never had in

* The "ratio" of two integers simply means
  one integer divided by another.
  For instance,
  the ratio of 3 to 2 is $\frac{3}{2}$.

Arithmetic.
These are the
NEGATIVE NUMBERS mentioned on page 9.
They are very easy to understand,
especially with the help of
the following line:

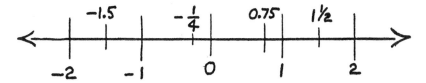

On this line you see that
all the numbers which
you had in Arithmetic are
the zero and the numbers to
the RIGHT of the zero,
including
integers, fractions,
decimal fractions, and mixed numbers,
each having its definite place on the line.
But you also see that we can put
numbers to the LEFT of the zero,
putting a MINUS sign in front of each
to distinguish it from
the numbers on the right,
called the POSITIVE numbers.
A positive number may be written
with a plus sign, thus:

$$+2, +5$$

or more briefly, without the plus sign:

$$2, 5.$$

But of course you must NEVER omit
the MINUS sign in front of
a negative number, like -1 or -7.

The NEGATIVE NUMBERS
are very USEFUL,
for they enable us to write
BRIEFLY:
a temperature of $5°$ BELOW ZERO
simply as $-5°$;
or a DEBT of 5 dollars
simply as $-\$5$.
Etc.
And as you go on,
you will see
other advantages of the
NEGATIVE NUMBERS.

So you see that
the first important thing about Algebra is
that it has already opened up to you
a new set of numbers
which do not exist in Arithmetic,
and which are very useful.

And the second new idea in Algebra is
the use of letters for numbers:

for instance,
in Arithmetic,
if you wish to find
the area of a rectangle
whose length is 4 feet and
whose width is 7 feet,
you say

$$\text{area} = 4 \times 7 = 28.$$

In Algebra,
you may use A for area,
l for length, and w for width,
and say

$$A = lw$$

which gives you
the general FORMULA for finding
the area of ANY rectangle,
instead of a particular one
as you do in Arithmetic.
You will find some more formulas on page 211.

Another advantage of
the use of letters is that,
in solving a problem,
you can do it more easily and
more directly
with the use of letters,
for you may use a letter to stand for
a quantity which is unknown,
and still go ahead and work with it,

and pretty soon
the problem almost solves itself
mechanically,
and you find out
the value of the unknown quantity!
It is almost like magic!

IV. THE RULES OF THE GAME

Now, in playing any game,
you naturally want to know
(1) what THINGS you play with—
that is, the EQUIPMENT, and
(2) what the RULES of the game are.
Think of baseball or football
or any other game
and you will see that
this is true.

It is the same way in
Science or Mathematics,
except that here we say
"ELEMENTS" instead of "EQUIPMENT,"
and the RULES are called
"POSTULATES."
Thus, in ARITHMETIC,
the ELEMENTS are the
POSITIVE RATIONAL NUMBERS AND ZERO;
and
the POSTULATES are very familiar to you,

although perhaps you never called them
"postulates" before.

Here are some of the postulates which
you know well:

(1) The Commutative Law for Addition:
that is, $a + b = b + a$,
which means that
when you ADD two numbers,
it does not matter which one
you write first:
for instance,

$$3 + 4 \text{ and } 4 + 3$$

both give the same answer, 7,
and therefore

$$3 + 4 = 4 + 3.$$

By the way,
notice how convenient it is to say

$$a + b = b + a$$

instead of merely

$$3 + 4 = 4 + 3,$$

because, when using letters,
you are really saying that
this Commutative* Law for Addition
applies to ALL the elements in
Arithmetic,

* Do you see why
  the word "commutative"
  is appropriate here?
  Do you know what a "commuter" is?

since the letters may stand for
ANY numbers,
and not merely for 3 and 4.

(2) The Commutative Law for Multiplication:
that is, $ab = ba$,
which means that
when you MULTIPLY two numbers,
it also does not matter which one
you write first:
for instance,

$$2 \times 5 = 5 \times 2.$$

Notice that with LETTERS
it is not necessary to write
the times sign ($\times$) between them,
as you do with numbers:
thus, ab means $a \times b$,
but 2×5 cannot be written 25,
since this means "twenty-five."

Perhaps you think that,
since ADDITION and MULTIPLICATION
are both COMMUTATIVE,
then all operations must be commutative.
But this is NOT so!
For instance,
DIVISION is NOT COMMUTATIVE,
since $12 \div 6 = 2$, but $6 \div 12 = \frac{1}{2}$;
or, in general,

$$a \div b \neq b \div a.*$$

Of course, all this is really
perfectly familiar to you,
for you learned it as a small child,
and now do it automatically.
Let us mention, therefore,
only a few more of
these familiar postulates,
which we shall need later.

(3) The Associative Law for Addition:

$$a + b + c = (a + b) + c = a + (b + c).$$

This means that
when you have to
ADD THREE NUMBERS together,
it does not matter whether
you add the sum of the first two
to the third,
or
add the first number to
the sum of the last two:
for instance,
to add $2 + 7 + 11$,
you may say either

$$2 + 7 = 9 \text{ and then } 9 + 11 = 20,$$

* Notice that instead of writing the words
"is not equal to,"
we simply write an "equal sign" (=) and
cross it out, like this $\neq$.

23

or you may say

2 + (the sum of 7 and 11, which is 18)

and then $2 + 18 = 20$;
you see that you get the same answer.
Notice that the parentheses
"associate"
the first two numbers in the one case,
and the last two numbers in the other:

$$(2 + 7) + 11 = 9 + 11 = 20,$$

or

$$2 + (7 + 11) = 2 + 18 = 20$$

and therefore

$$(2 + 7) + 11 = 2 + (7 + 11).$$

Or, in general,

$$(a + b) + c = a + (b + c),$$

that is why this is called the
ASSOCIATIVE LAW.

(4) The Associative Law for Multiplication:

$$abc = (ab)c = a(bc).$$

And
(5) The Distributive Law:

$$a(b + c) = ab + ac.$$

An illustration of this is:

if you wish to
MULTIPLY a number, like 5,
by the SUM of two other numbers,
say, $2 + 7$,
you may say either

$$5 \times (2 + 7) = 5 \times 9 = 45$$

or you may multiply 5×2 and 5×7,
and add these results together, obtaining

$$5 \times (2 + 7) = 5 \times 2 + 5 \times 7 = 10 + 35$$

which gives the same answer, 45.
Thus $5(x + y)$ is the same as $5x + 5y$
or $7c + 7d$ is the same as $7(c + d)$.

When you say that $5(x + y)$
is equal to $5x + 5y$,
you are doing
a MULTIPLICATION example.
But when you say that $5x + 5y$
is equal to $5(x + y)$
you are FACTORING;
just as $3 \times 2 = 6$
is a MULTIPLICATION example
but when you say

$$6 = 3 \times 2$$

you have split 6 into
its two FACTORS, 3 and 2.
Note that when you

split 6 into $5 + 1$
you have NOT factored the 6.
To FACTOR a quantity
you must split it into
a PRODUCT of other quantities.
Thus, to factor 14
you get 7×2;
and to factor $3x + 9y$
you get $3(x + 3y)$,
etc.

To be sure that you understand this
you should practise a little
on page 191 #1.
It is fun!

As we have already told you,
in Algebra we have also
NEGATIVE NUMBERS (see page 13),
so that here the elements will be
ALL THE RATIONAL NUMBERS,*
POSITIVE and NEGATIVE, and ZERO.
And,
although we have introduced
new things to "play with"
(the negative numbers),

* As you continue your study of Algebra,
  we shall introduce you also to
  other numbers,
  called IRRATIONAL and IMAGINARY NUMBERS!
  But you do not have to worry about that yet.

still,
we do not change any of the
above-mentioned RULES of the game
(the POSTULATES*),
since, after all,
we are still playing
practically the same game,
Algebra being only a sort of
"glorified" Arithmetic.

* The postulates mentioned here are
  not a COMPLETE set of postulates for Algebra.
  If some time you should want to see
  a complete set,
  you will find it in a little book by
  Professor E. V. Huntington of Harvard University,
  called
  The Fundamental Propositions of Algebra,
  published by
  The Galois Institute Press of
  Long Island University, Brooklyn, N.Y.

V. HOW THE GAME IS PLAYED

And now let us see
how the game is played.
First, take Arithmetic:
As soon as you learned the
integers
and the postulates (the rules),
you were then taught
how to perform the
FOUR FUNDAMENTAL OPERATIONS
(ADDITION, SUBTRACTION,
MULTIPLICATION, DIVISION)
with the integers.
And later,
when you had fractions,
you again had to learn
how to perform these
FOUR FUNDAMENTAL OPERATIONS
with fractions:
that is,
how to ADD FRACTIONS,
how to SUBTRACT them,

how to MULTIPLY them,
and how to DIVIDE them.
(And, similarly
with DECIMAL fractions
and mixed numbers.)
So that, by now
you should know how to perform the
FOUR FUNDAMENTAL OPERATIONS
with all the
POSITIVE RATIONAL NUMBERS.
But in case you have forgotten this,
and since it is very important,
you had better take a little time out
to review this.
Try your skill on page 191 #2.

And now
all you have to learn in Algebra is
how to perform the
FOUR FUNDAMENTAL OPERATIONS
with NEGATIVE NUMBERS
and with LETTERS.
Then you can apply it to
the solution of
practical problems

VI. ADDITION

Now you know that in Arithmetic,

$$7 + 4 = 11;$$

and, of course, this is still true
in Algebra.
But in Algebra you might have
an example like this:

$$7 + (-4).$$

What does this mean?
Well, if you are talking about money,
a positive number (like 7) represents
money that you HAVE, or
that is COMING IN,
whereas
a negative number (like -4) represents
money that you OWE, or
that is GOING OUT
(like expenses).
And so,
if you have $7

and a debt of $4,
your account is worth $3,
so that

$$7 + (-4) = 3.$$

In other words, you may think of
ALGEBRAIC ADDITION as
"balancing an account."
See if you understand
the answers to
the following examples:

(1) Add: (2) Add: (3) Add: (4) Add:
   4   4   −4   −4
   7   −7   7   −7
  11 Ans.   −3 Ans.   3 Ans.   −11 Ans.

(5) Add: (6) Add: (7) Add:
   −3   1   −3
    7   0   −5
   11   −1   −6
   −5   4
   10 Ans.   4 Ans.   −14 Ans.

These examples may also be written
horizontally, thus:

(1) $4 + 7 = 11$; (2) $4 - 7 = -3$; (3) $-4 + 7 = 3$;
(4) $-4 - 7 = -11$; (5) $-3 + 7 + 11 - 5 = 10$;
(6) $1 + 0 - 1 + 4 = 4$; (7) $-3 - 5 - 6 = -14$

At first, in doing such examples,
it might be easier for you
to have a little "Account Book,"
and put on the RIGHT-HAND page
the heading "Coming In,"
and on the LEFT-hand page
the heading "Going Out";
then,
in doing these "addition" examples,
put all POSITIVE numbers on the RIGHT-hand page,
and all NEGATIVE numbers on the LEFT-hand page,
and then "balance" your account.
Thus, in example (5) above,
7 and 11 would go on the RIGHT,
−3 and −5 on the LEFT,
then, in "balancing"
you see that
$18 are "coming in,"
$8 are "going out,"
so that the result is that
you would have $10 left over.
Whereas, in example (7),
since all the items are "going out,"
the result is
a debt of 14,
or −14.
Notice that the meaning of
the word "add"
is NOT exactly the same as in
ARITHMETIC,

but had to be modified so that
we can apply it to
NEGATIVE numbers;
and yet, when the numbers
happen to be all POSITIVE,
(as in example (1) above),
the answer IS exactly the same as in
Arithmetic,
so that the
new definition of "addition"
DOES NOT CONTRADICT the old one,
but merely enlarges it
to make it applicable
to the new elements
(the negative numbers).

You will see,
as you go on in Mathematics,
that,
to MAKE PROGRESS POSSIBLE,

(1) Old definitions have to be modified,
(2) New elements have to be introduced,* and
(3) Even new postulates have to be introduced.

Thus the old ideas are
NOT ENTIRELY discarded,
but are MODIFIED as

* See page 26.

the need arises.
This is absolutely essential
to MAKE PROGRESS POSSIBLE!

So you see that
in Mathematics
we do not say,
"What was good enough for
my grandfather
is good enough for me,"
but modify our ideas to
suit the times we live in.

Notice that
the MINUS sign in front of the 7
in example (2) on page 32
does NOT mean "subtract,"
for this is an ADDITION example;
it merely says that the 7 is
A NEGATIVE number.
And you will soon see that
you can do all sorts of things with
negative numbers,
you can ADD THEM, SUBTRACT THEM,
MULTIPLY THEM, and DIVIDE THEM;
so that from now on
you must NOT think that
every minus sign means "subtract,"
as it did in Arithmetic.

And now how about adding LETTERS?*
Well, this too is quite easy.
For, if you wish to add
2a and 9a, you get 11a,
or 2a + 9a = 11a.

Also:

(1) Add:	(2) Add:	(3) Add:
2a	−2x	−2ab
−9a	9x	−9ab
−7a Ans.	7x Ans.	−11ab Ans.

And this is true no matter what
the a, x, or b stands for,
for it is only the
"numerical COEFFICIENTS"
(the numbers in front of the letters)
which tell you
how MANY of the letters you have,
whereas the letters are merely
the NAMES of the things you are adding.
Thus if x stands for the number "seven,"
then obviously
"2 sevens + 9 sevens" gives "11 sevens,"
or,
if x stands for the number "five,"
then again
"2 fives + 9 fives" gives "11 fives,"

*But before you go on with this
you can have a little fun with yourself:
see page 192, #3.

etc.

And of course the signs are treated
in the same way
whether there are any letters or not
(as you can see in the three examples on page 37).
And, if you have a letter (or letters)
WITHOUT ANY NUMERICAL COEFFICIENT,
like a, x, or abc,
this naturally means

$$1a, \ 1x, \ 1abc.$$

Hence the following answers:

(1) Add:

$$\begin{array}{r} xy \\ -3xy \\ \hline -2xy \text{ Ans.} \end{array}$$

(2) Add:

$$\begin{array}{r} -4abc \\ -abc \\ +5abc \\ \hline 0 \quad \text{Ans.} \end{array}$$

(3) $-8a - a + 7a + 5a - 9a = -6a$ Ans.

Obviously, in order to be able
to add at all,
the KIND of things must be the same;
that is,
you can add $2a$ and $5a$,
(and get $7a$),
but NOT $2a$ and $5b$:
In the latter case,
you merely write
$2a + 5b$ for your answer.

And now you will enjoy
some examples to practise on again (page 192 #4).

VII. SUBTRACTION

Next we come to another of the
FOUR FUNDAMENTAL OPERATIONS,
SUBTRACTION,
and you must learn how to do
subtraction with
NEGATIVE NUMBERS and LETTERS.

In the first place,
if you stop to realize that the word
"subtract" means
"take away,"
you will easily understand
what it means to
"subtract" or
"take away" or
"erase"
an item from your little "Account Book."*
Thus to subtract $7 from your account,
you would have to
erase $7 from the RIGHT-hand side;

*See page 33.

and to subtract $-7 (or a DEBT of $7),
you would have to
erase an item of $-7 from the LEFT-hand side,
since all DEBTS were entered on
the LEFT-hand page.
But perhaps you are
a very neat person,
and do not like to
mess up your little book by
erasing items from it.
In that case we can show you
a very neat trick:

Instead of
ERASING $7 from the RIGHT-hand page,
you can
ENTER $-7 on the LEFT,
for such an entry
naturally cancels out
the $7 on the right,
does it not?
Give this a little thought,
for it is a very
useful and important idea.
And, similarly,
if you wish to
ERASE an item of $-7 from the LEFT,
you may INSTEAD
ENTER an item of $7 on the RIGHT
for such an entry has the effect

of canceling out
or eliminating
a DEBT of $—7 on the LEFT.

Of course,
after making the entry,
you balance your account
as before.
In short,
every SUBTRACTION example
may be changed to
an ADDITION example
by the simple device
mentioned above:
so that you do not have to worry
about subtraction at all!

Let us show you how it works.
(1) Subtract:

$$\begin{array}{r} 7 \\ 4 \\ \hline 3 \text{ Ans.} \end{array}$$

This means that your little book
may look like this at first,
showing that your account is
originally worth $7:

Out	In
−6	· 5
−4	4
−1	9

and you wish to TAKE AWAY $4.
You can of course merely cross out
the $4 like this:

Out	In
−6	5
−4	4̶
−1	9

And your account is now worth
only $3, is it not?
But, if you do not like to
cross out the $4,
you may INSTEAD
ENTER an item of $−4 on the LEFT,
like this:

Out	In
−6	5
−4	4
−1	9
−4	

and, if you now balance THIS account,
you see that it is worth $3,
just as before.
So that
CROSSING OUT $4 ON THE RIGHT
has the SAME effect as
ENTERING $−4 ON THE LEFT!
Which is another way of saying that
SUBTRACTING $4
has exactly the SAME effect as
ADDING $−4.
Hence, the following two examples
have the SAME answer:

(1) Subtract:

$$\begin{array}{r} 7 \\ \underline{4} \\ 3 \end{array}$$

(2) Add:

$$\begin{array}{r} 7 \\ \underline{-4} \\ 3 \end{array}$$

Similarly,
if you wish to
TAKE AWAY $−4 FROM THE LEFT,

you may INSTEAD
ADD $4 ON THE RIGHT, thus:

Out	In	This account
−6	5	is now
−4	4	worth $7,
−1	9	is it not?

If you now cross out $−4, thus:

Out	In	This account
−6	5	is now
−4	4	worth $11,
−1	9	is it not?

Or ENTER $4 on the right instead, thus:

Out	In	This account
−6	5	is now also
−4	4	worth $11,
−1	9	is it not?
	4	

So that
taking away or crossing out or
SUBTRACTING the $-4
has the SAME effect as
entering or
ADDING the $4 on the RIGHT!
Hence, the following two examples
have the SAME answer:

$$(1)\ \text{Subtract:} \qquad (2)\ \text{Add:}$$

$$
\begin{array}{cc}
7 & 7 \\
\underline{-4} & \underline{4} \\
11 & 11
\end{array}
$$

Try a few more similar examples,
until you are
thoroughly convinced that:
to SUBTRACT a POSITIVE number
you may instead
"ADD" a NEGATIVE number;
and
to SUBTRACT a NEGATIVE number
you may instead
"ADD" a POSITIVE number.
Or, to put it more BRIEFLY:
to SUBTRACT ANY NUMBER,
you may instead
ADD ITS INVERSE.
Perhaps you can guess the meaning
of this word "INVERSE":
the inverse of 3 is -3,

Hold it, small fry!
DON'T SUBTRACT a number by
erasing it! Just ADD its
INVERSE on the opposite page!
You will get the SAME answer —
TRY IT and SEE!

and the inverse of -3 is 3,
or, in general,
x and $-x$ are
inverses of each other.*

If you are convinced
that you have the right to do this,
you will find it very easy
to do the following examples:
(1) Subtract:

$$\begin{array}{r} -9 \\ -2 \\ \hline \end{array}$$

Since this is equivalent to:
(2) Add:

$$\begin{array}{r} -9 \\ 2 \\ \hline \end{array}$$

therefore the answer is -7.
Notice that it is always the
LOWER number whose sign
you must change,
since that is the item
which is to be subtracted or
"erased."

Now see if you understand
the following answers:

* Later on, the word "INVERSE" will have
  a broader meaning,
  but, for the time being, this will do.

(3) Subtract:

$$9a$$
$$-2a$$
$$\overline{11a} \text{ Ans.}$$

(4) Subtract:

$$-9xy$$
$$2xy$$
$$\overline{-11xy} \text{ Ans.}$$

At first, until you get used to it,
change each of these
to the corresponding addition example,
but as soon as possible
learn to make the change
MENTALLY,
and write down your answer
immediately.

And now again,
bring on those exercises
for PRACTICE, on page 193 #5.

VIII. RAPID PROGRESS

Remember that
in order to be able
to ADD at all,
the KIND of things must be the same
(see page 38);
that is,
you can add $7ab$ and $4ab$ and get $11ab$,
but you cannot add $7ab$ and $4xy$;
in this latter case,
you merely write

$$7ab + 4xy$$

for your answer.
And, similarly,
to add $-4xy$ to $7ab$,
you merely write for your answer:

$$-4xy + 7ab$$

or

$$7ab - 4xy.$$

An expression like $7ab + 4xy$
is said to have TWO TERMS in it:

$7ab$ is the first TERM and
$4xy$ is the second TERM
and in $7ab - 4xy$,
$7ab$ is the first TERM and
$-4xy$ is the second one.
And
$-4a + 7bc - 9ab$
has THREE TERMS:
$-4a$, $7bc$, and $-9ab$.

Any expression having
ONE OR MORE TERMS IN IT
is called a
POLYNOMIAL*;
and, more specifically,
a polynomial of
exactly ONE term is a
MONOMIAL;
if it has exactly TWO terms,
it may be called a
BINOMIAL;
a THREE-TERM expression is a
TRINOMIAL.
Notice that
TERMS ARE SEPARATED BY
$+$ or $-$,
BUT NOT BY

* Later we shall give you
  a better definition of "polynomial,"
  but this will do for the present.

$\times$ or $\div$.

Thus $7 \times a \times b \div c$ is
ONLY ONE TERM.

Similarly,

$11ab$ is ONLY ONE TERM,
since times signs are understood
between the 11 and the a,
and between the a and the b (see page 22.)

Finally,
consider the following examples:

(1) Add:

$$2a + 3b - 7ac$$
$$\underline{-4a + \ b - \ ac}$$
$$-2a + 4b - 8ac \text{ Ans.*}$$

(2) Add:

$$-5xy - \ ax + 7$$
$$ax - 8 + 4d$$
$$\underline{7xy + 2ax + 1 - 9d}$$
$$2xy + 2ax \qquad - 5d \text{ Ans}$$

You see that
if you must add various things,
you put "like" or "similar" terms
under each other,
and then
add each column;
and you see that
"similar" terms are terms which
have the same letters in them.
(Later, after you have had "exponents,"

* Example (1) may of course be written
horizontally, thus:
$2a + 3b - 7ac - 4a + b - ac = -2a + 4b - 8ac$
(see page 32).
And similarly for example (2).

You will get a better definition of
"similar terms,"
but this will do for the present.)

Also in subtraction
arrange similar terms in columns
and subtract in each column, thus:
(1) Subtract $2x - 7ab$ from $7x + 2ab - 5$.
   Arrange in columns and
   ADD the INVERSE of
   the subtrahend (the LOWER line) to
   The minuend (the UPPER line)
   as explained on page 45:

$$7x + 2ab - 5$$
$$\underline{2x - 7ab}$$
$$5x + 9ab - 5 \text{ Ans.}$$

(2) From $9a - 5$ subtract $2a + 7b$.
   Again, arrange in columns and
   add the INVERSE of the subtrahend
   to the minuend:

$$9a - 5$$
$$\underline{2a \qquad + 7b}$$
$$7a - 5 - 7b \text{ Ans.}$$

Of course you must be
VERY CAREFUL to know
which expression should be written
on the LOWER line,
or you will be taking
the inverse of the wrong one!

Just one more thing
to show you
the value of
negative numbers.
In Arithmetic, as you know,
where there are
NO negative numbers,
you cannot subtract a
LARGER number from a SMALLER one:
you cannot take 9 from 7,
you just say
it is impossible!
But, in Algebra,
it is quite EASY
to take 9 from 7,
obtaining -2 for the answer.
And this is a very PRACTICAL situation,
for many a person
buys $9 worth of goods when
he has only $7,
if his credit is good!
In fact,
the entire business world
depends upon credit.
So you see that, in Algebra,
subtraction is
NEVER IMPOSSIBLE!

Time out for practice again,
on page 193 #6 and #7.

IX. SOME APPLICATIONS

You undoubtedly know that
a triangle is a figure bounded by
three straight lines and having
three angles
(that is why it is called a *tri*angle)
like this:

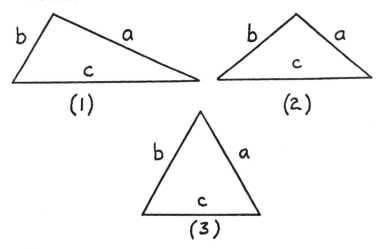

There are of course
triangles of various shapes and sizes:

In (1) the three sides have different lengths.

In (2) side b = side a:

such a triangle,
having two sides equal,
is called an "isosceles triangle."

In (3) $a = b = c$:

such a triangle is called
an "equilateral triangle."

Suppose that in (2),
side a is $2x + 7y$ inches long.
How long is side b?

In (3), if b is $5m - 6z$ inches long,
how long are the other two sides?
How long is the "perimeter" in (3),
that is,
the sum of all three sides?
If p represents the perimeter,
what does the following formula say?

$$p = a + b + c$$

Translate it into plain English.*

If in triangle (1),

> side $a = 11xy - 4$ inches,
> side $b = 9x - 5$ inches,
> side $c = xy - 1$ inches,

find its perimeter.
How much longer is side a than side b?*

As we said before,
a triangle has also three angles.
An angle, as you know,
is the opening between
two lines which meet at a point,
like this:

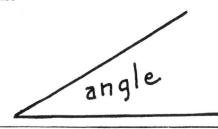

and of being
INTERNATIONAL!
If you write in
Algebra language,
you will be understood
not only in
your own country
but also in
Russia,
France,
and all over!

* You will find the right answers
on page 220,
but do not look too soon!
Give yourself a chance first.

To increase the angle,
you must spread the lines further apart,
thus:

If you merely lengthen the lines
without spreading them,
this does NOT change
the SIZE of the angle:

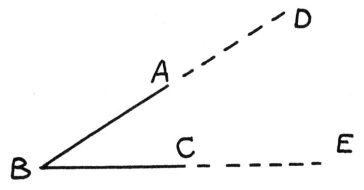

Here, angle *ABC* is exactly
The SAME size as angle *DBE*.

Angles may be measured in degrees.
If you form two angles of
exactly the same size, like this:

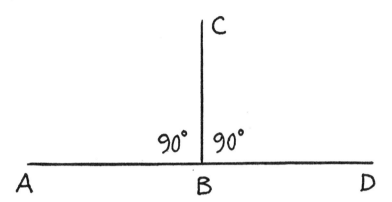

Where *ABD* is a continuous straight line,
each of these is a "right angle"
and has 90 degrees in it.
An angle less than a right angle is
an "acute angle";
an angle greater than a right angle is
an "obtuse angle":

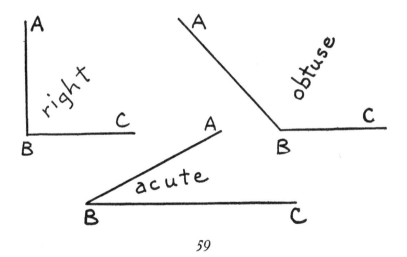

The point B, where the two lines meet,
is called the "vertex" of the angle.
The lines themselves are called the
"sides" of the angle:
BA and BC are the sides.
You can of course spread the lines apart
even more than in an obtuse angle;
when you spread them so that
the sides are in the same straight line,
you get a "straight angle," like this:

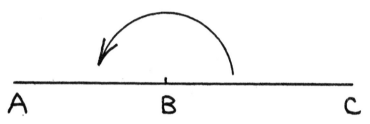

A   B   C

The vertex is still at B;
the sides are still BA and BC;
and the angle has 180 degrees in it.
(Do you see that it is really
twice as large as a right angle?)
Notice that this is NOT the same
as a "straight line";
the straight line is AC, and
it has NO vertex;
whereas the straight angle is
the *opening* between the sides BA and BC.
If you swing the side BA still further,
you can have angles larger than 180 degrees:

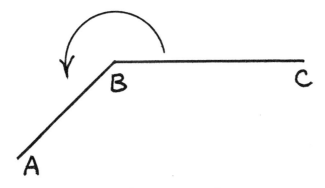

If you swing *BA* all the way around
until it gets to *BC*,
you would have an angle of 360 degrees,
and if you go still further,
you can have angles of 480 and 720 degrees, and so on,
without limit!
When a dancer
pirouettes 25 times,
the angle through which she has turned
is 25 times 360 degrees
or 9000 degrees,
and expert dancers can turn through
even more than this.
So you see that angles can be
of unlimited size.

If the sum of TWO angles is 90 degrees,
they are called
COMPLEMENTARY ANGLES,
and each is the COMPLEMENT of the other.

GEE, do you think she's turned through a million degrees yet?

Oh, you're beginning to see math even in a movie!
You know somethin'? Without math you wouldn't even have anything to eat or wear or a bed to sleep in!

If the sum of TWO angles is 180 degrees,
they are called
SUPPLEMENTARY ANGLES,
and each is the SUPPLEMENT of the other.
DO NOT CONFUSE THESE WORDS!
Thus if an angle has 30 degrees in it,
its COMPLEMENT is $90° - 30°$ or $60°$;
and its SUPPLEMENT is $180° - 30°$ or $150°$.
Similarly, if an angle has x degrees,
its complement is simply $90° - x°$:
you cannot combine these two terms into one,
as you did with the
two numbers, $90° - 30° = 60°$,
since in $90 - x$, the two terms are
NOT SIMILAR (see page 51);
also,
the supplement of x degrees is $180 - x$.

Suppose an angle has $4xy - 5m$ degrees;
to find its supplement you would have to
subtract this from 180 degrees,
like this:
Subtract:

$$\begin{array}{r} 180 \\ \underline{4xy - 5m} \\ 180 - 4xy + 5m \text{ Ans.} \end{array}$$

Notice that we use separate columns
for each kind of term,
and then

ADD the INVERSE of the LOWER quantity
to the top line (see page 52).

Now try the following exercises:

(1) If an angle has $7x - 2y$ degrees,
find its complement.

(2) If one angle has $11x + 45$ degrees,
and another has $-11x + 45$ degrees,
show that these angles are
complementary.

(3) Suppose one angle has x degrees and
another has y degrees.
If they are supplementary, then

$$x + y = 180 \text{ (see page 63).}$$

Do you understand this?

Now suppose one angle has m degrees, and
Another has 30 degrees.
If they are complementary,
how would you express this fact
in algebraic "shorthand," as above?

And now try
some more applications of
Algebraic ADDITION and SUBTRACTION
on page 194 #8.

X. MULTIPLICATION AND DIVISION

And now you have only two more of
the FOUR FUNDAMENTAL OPERATIONS
to learn how to perform—
namely,
how to MULTIPLY and to DIVIDE with
NEGATIVE NUMBERS and LETTERS.

Let us again go back to
Arithmetic,
for a moment.
You know that for integers
the multiplication example, 3 × 2,
may be regarded as a short way of doing
the following ADDITION example:

$$\begin{array}{r} 2 \\ 2 \\ \underline{2} \\ 6 \text{ Ans.} \end{array}$$

That is,
writing the 2 THREE TIMES,

and ADDING;
and you learned to do this
more BRIEFLY thus:

$$3 \times 2 = 6.$$

Similarly,

$$3 \times (-2)$$

may be written:
Add:

$$-2$$
$$-2$$
$$-2$$
$$\overline{-6} \text{ Ans.}$$

And therefore,
more BRIEFLY

$$3 \times (-2) = -6.$$

What then are the answers to
the following exercises?

(a) $7 \times (-9)$ (c) $5 \times (-2)$
(b) $8 \times (-1)$ (d) $11 \times (-7)$

This is quite simple, is it not?*
But how about this one?

$$-3 \times 2.$$

Here you are asked to find

$$-3 \text{ times } 2,$$

but

* Don't look now but
  the answers are on page 220.

you CANNOT write 2, MINUS 3 times,
can you?
And so you must figure it out
some other way,
like this:
Instead of trying to do

$$-3 \times 2,$$

you may do instead

$$2 \times (-3),$$

since
MULTIPLICATION IS COMMUTATIVE
(see page 22);
and now you CAN write
the -3, TWO TIMES, thus:

$$\begin{array}{r} -3 \\ \underline{-3} \end{array}$$

And, adding as before,
you get the answer, -6.
So now you know how to multiply:

[1] A POSITIVE integer by
another POSITIVE one,
the answer being also
a POSITIVE integer,
like $3 \times 2 = 6$.

[2] A POSITIVE integer by
a NEGATIVE one,

the answer being
a NEGATIVE integer,
as in $3 \times (-2) = -6$.

[3] A NEGATIVE integer by
a POSITIVE one,
the answer being
a NEGATIVE integer,
as in $(-3) \times 2 = -6$
(explained on page 67).
And now there is only
one more possible case,
namely,
[4] How to multiply
a NEGATIVE integer by
another NEGATIVE one,
like
$$(-3) \times (-2).$$
In this case you see that the
Commutative Law of Multiplication
does not help us here
as it did in case [3],
since
you cannot write any number
MINUS TWO TIMES,
any better than you can write any number
MINUS THREE TIMES,
so you are stuck either way!
Now let us see what
we can do about it.

Consider the following trinomial:

$$ab + a(-b) + (-a)(-b)$$

where a and b are any two integers;
we may think of this in
two different ways:

(1) The Distributive Law (page 24)
permits us to write
the LAST TWO TERMS like this:

$$(-b)[a + (-a)],$$

does it not?
But $a + (-a)$ equals zero,
and any number times zero is
also zero,
making $(-b) \times 0 = 0$,
so that the value of the entire trinomial
becomes ab,
since the *last two* terms give zero.
Or

(2) The Distributive Law also permits us
to write the FIRST TWO TERMS
like this:

$$a[b + (-b)];$$

here again

$$b + (-b) = 0,$$

and $a \times 0 = 0$;
therefore, the value of the trinomial
becomes $(-a)(-b)$,
since the *first two* terms give zero.

And consequently,
putting these two ways together,
since the Associative Law for Addition
(see page 23)
permits us to evaluate our trinomial EITHER way,
we conclude that
$(-a)(-b)$ must equal ab, since
They are BOTH equal to the SAME thing,
namely,
the value of the trinomial on page 69.
Well,
do you realize what we have
just PROVED?
Do you understand what

$$(-a)(-b) = ab$$

tells you?
Why obviously it tells you that
when you multiply
TWO NEGATIVE NUMBERS
you get the same answer as if
the two integers were
both POSITIVE!
Hence

$$(-3)(-2) = 3 \times 2 = 6.$$

And you cannot possibly deny this
if you agreed to
the simple postulates which
we gave you in Chapter IV,
and which are some of

the RULES of this game.
Probably in the beginning
the postulates seemed to you
so reasonable that you thought they were
hardly worth mentioning,
but you see
you can do some nice tricks with them!*

And now let us give you
a VERY SHORT RULE
summarizing all four cases,
[1], [2], [3], [4], given above,
so that it will be easy for you to
multiply ANY two integers:

IN MULTIPLICATION
TWO LIKE SIGNS GIVE PLUS,
and
TWO UNLIKE SIGNS GIVE MINUS.

This means that if the signs are
BOTH PLUS or BOTH MINUS,

* Some day, if you go on with your study of
  Mathematics,
  you will find
  other Arithmetics and Algebras in which
  the postulates are entirely different
  from those so familiar to you,
  and you will be surprised to see
  how interesting and useful
  those funny Arithmetics are!
  If you can't wait till you get a little older,
  you can read about these in
  The Education of T. C. Mits by L. R. and H. G. Lieber
  (published by Norton & Co.).

the answer will be PLUS;
and
if EITHER sign is PLUS and the other MINUS,
then the answer will be MINUS.

Try this rule on the following exercises:

(a) $5 \times (-7)$ (b) 5×7
(c) $(-5) \times (-7)$ (d) $(-5) \times 7.$

Did you get these answers?

(a) -35 (b) 35
(c) 35 (d) $-35.$

Then you are 100 per cent right.

And now we have very good news for you,
namely, that
the RULE for SIGNS in DIVISION
is EXACTLY THE SAME as
in MULTIPLICATION!

Let us show you why:

Since $3 \times 2 = 6$
then $6 \div 3 = 2,$
and $6 \div 2 = 3;$
in other words,
if you multiply two integers,
then the product (in this case, 6)
when divided by
either of the two original integers

of course gives the other one
as a result of the division,
does it not?

Well, applying this idea, we get:

since $3 \times 2 = 6$, then
(a) $6 \div 3 = 2$;

since $3 \times (-2) = -6$, then
(b) $-6 \div 3 = -2$;

since $3 \times (-2) = -6$, then
(c) $(-6) \div (-2) = 3$;

since $(-3) \times (-2) = 6$, then
(d) $6 \div (-3) = -2$.

In other words,
when you are dividing one integer by another,
if BOTH signs are ALIKE—
BOTH PLUS as in (a) or
BOTH MINUS as in (c)—
the answer is PLUS;
and if the two SIGNS are NOT ALIKE—
one PLUS and the other MINUS,
as in (b) and (d)—
then the answer is MINUS.
Or, more BRIEFLY,
TWO LIKE SIGNS GIVE PLUS and
TWO UNLIKE SIGNS GIVE MINUS,
just as in Multiplication.

But all this is about
INTEGERS.
How about the
rule of signs for
multiplying and dividing
OTHER RATIONAL NUMBERS,
like

$$\tfrac{3}{2} \times (\tfrac{7}{8})?$$

Well, fortunately, it is
STILL THE SAME RULE:
TWO LIKE SIGNS GIVE PLUS and
TWO UNLIKE SIGNS GIVE MINUS,
as you can easily see from
the following argument:

Since $\dfrac{-12}{3} = -4$ (see page 73),

and $-\left(\dfrac{12}{3}\right)$ also equals -4,

then $-\left(\dfrac{12}{3}\right)$ is the same as $\dfrac{-12}{3}$,

since they both equal -4.
Or, in general, for ANY integers a and b,

$$\left(\dfrac{a}{b}\right) = \dfrac{-a}{b}$$

therefore,
if you wish to multiply

$$\dfrac{3}{2} \times \left(-\dfrac{7}{8}\right)$$

you may write this

$$\frac{3}{2} \times \frac{-7}{8}$$

and, multiplying these fractions
in the usual way*
we get

$$3 \times (-7) = -21$$

and

$$2 \times 8 = 16,$$

giving the answer $\dfrac{-21}{16}$ or $-\dfrac{21}{16}$;
in other words,
PLUS times MINUS gives MINUS, as before.

Taking another illustration,

$$(-\tfrac{8}{9}) \times (-\tfrac{2}{5}),$$

we may write this

$$\frac{-8}{9} \times \frac{-2}{5}$$

and therefore the answer is $\tfrac{16}{45}$,
so that
MINUS times MINUS gives PLUS,
as before,
etc.

And similarly for Division.

* That is,
  to multiply two fractions,
  you must multiply the two numerators,
  and multiply the two denominators,
  as shown above.

Of course if you have to multiply
more than two numbers,
you do two at a time, like this:
$(-2) \times (6) \times (-3) = (-12) \times (-3) = 36$ Ans.
And now go and get yourself some
PRACTICE on page 195 #9.

XI. MULTIPLICATION AND DIVISION
WITH LETTERS

Now that you know how to perform the
FOUR FUNDAMENTAL OPERATIONS with
POSITIVE and NEGATIVE
RATIONAL NUMBERS,
and
since you have already seen how
to ADD and SUBTRACT with LETTERS
(see pages 51, 52)
all you have to learn now is how to
MULTIPLY and DIVIDE with LETTERS—
and this too is very simple.

For example,
you already know from Arithmetic that
7×7 may be written 7^2;
that is,
when you have to multiply
the SAME number several times,
you may indicate it by
writing the number to be multiplied

with a SMALL number above it on the RIGHT,
which shows HOW MANY times
it is to be used as a
multiplying factor.
The first number (7), is called the
BASE,
and the LITTLE number (2) is called the
EXPONENT.
Thus, in 5^3,
5 is the base and
3 is the exponent, and
this is a short way of writing

$$5 \times 5 \times 5,$$

taking the 5, THREE times, and
MULTIPLYING them,
giving the answer 125,
so that $5^3 = 125$.
NOTICE THAT
5^3 does NOT mean $5 \times 3 = 15$;
be VERY CAREFUL about this!

To read 5^4, you say
"5, exponent 4";
a^m is read "a, exponent m";
when the exponent happens to be 2,
you read it "square," like this,
7^2 is read "7 square";
when the exponent is 3, you say "cube," thus:
5^3 is "5 cube."

Otherwise, just say
2^5 is "2, exponent 5,"
etc.

Now try reading and evaluating these:

$$7^2, \qquad 3^3, \qquad 2^5, \qquad 4^3.$$

Which of the following answers are right?

$$49, \qquad 9, \qquad 10, \qquad 64$$

When there is NO exponent written,
as in 7,
the exponent 1 is of course understood,
since we have here just one seven,
and not, say, two sevens as in 7^2,
or three sevens as in 7^3;
but REMEMBER that
the two sevens in 7^2 are to be
MULTIPLIED, so that

$$7^2 = 7 \times 7 = 49$$

and NOT $7 \times 2 = 14$,
as we have already warned you above.

Now if we have a letter, say, b,
you know that it stands for a number;
And therefore

$$b^2 = b \times b$$

and

$$c^3 = c \times c \times c$$

just exactly as with numbers,
since the letters really ARE numbers.

And now suppose you wish to multiply

$$a^3 \times a^2.$$

Since $a^3 = a \times a \times a$
and $a^2 = a \times a$
therefore

$$a^3 \times a^2 = a \times a \times a \times a \times a$$

which of course may be written

$$a^5$$

since there are
FIVE a's all MULTIPLIED together.
Similarly

$$b^4 \times b^3 = b \times b \times b \times b \times b \times b \times b = b^7$$

Now you can easily see that
you do not have to go to the trouble of
writing out all the separate b's;
since you can easily see that $b^4 \times b^3$
will give you SEVEN b's
ALL MULTIPLIED together.
And hence when you wish to MULTIPLY
a base with an exponent by the
SAME base with another exponent,
you get the answer by the following
simple rule for MULTIPLICATION:
RETAIN THE BASE AND
ADD THE EXPONENTS.

Do you clearly see why
you ADD the exponents although it is
a MULTIPLICATION example?
You notice that $b^4 \times b^3 = b^7$ (see above),
and that is simply because
there are SEVEN b's all
MULTIPLIED together,
so that ADDING the 4 and 3
is merely for the purpose of
finding out
HOW MANY b's there are that have to be
MULTIPLIED together.

What are the correct answers to these?

$$c^3 \cdot c^5, \qquad m^7 \cdot m, \qquad a^4 \cdot a^2, \qquad y^m \cdot y^s$$

In the second, DO NOT FORGET that
m means m^1 (see page 80),
and that a times sign is understood
between letters in
all four examples.

In the fourth, did you get y^{m+s}?
If not,
read again the rule on page 81,
which says
RETAIN the BASE (y in this case) and
ADD the EXPONENTS
(hence $m + s$ is the correct exponent in the answer).

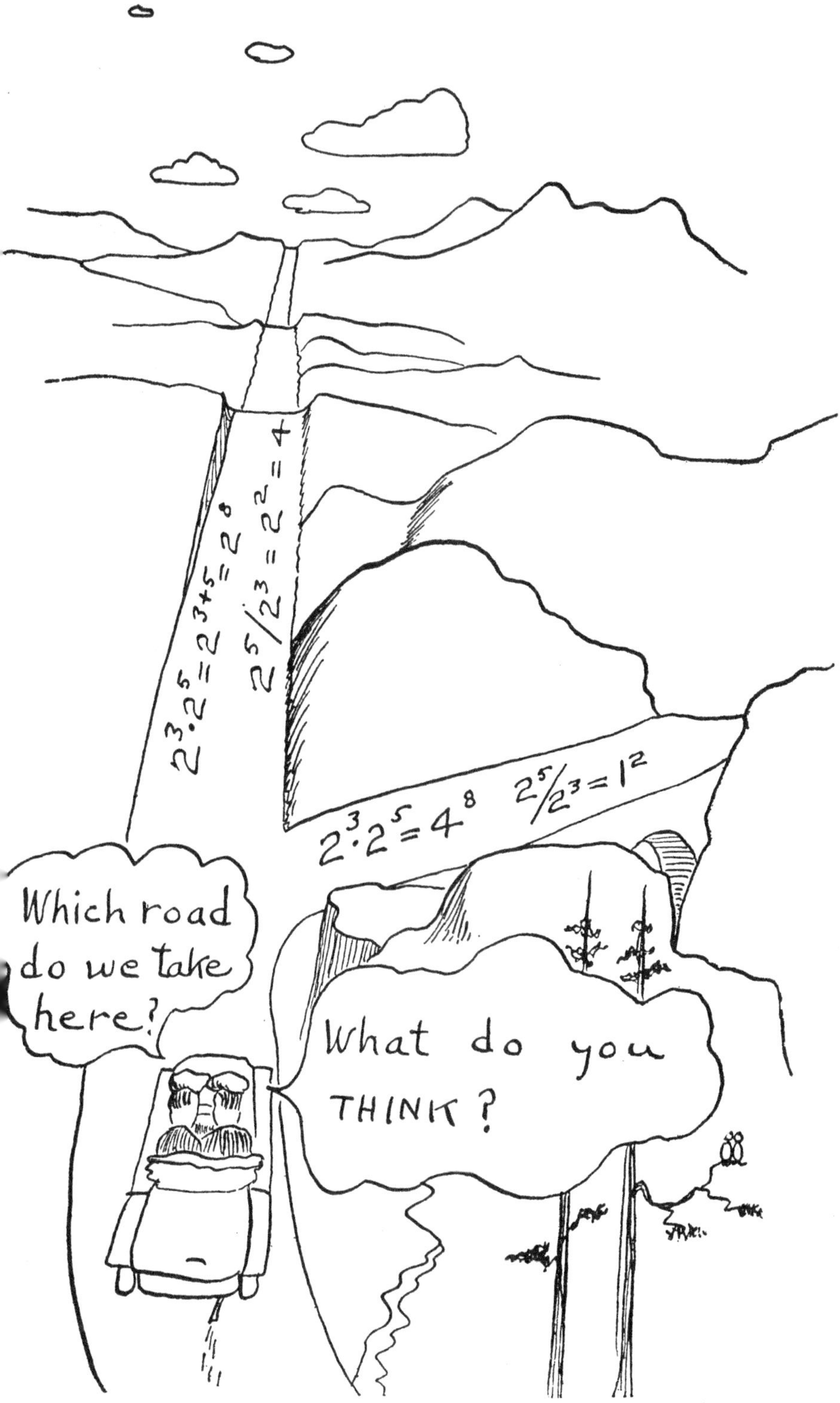
$2^3 \cdot 2^5 = 2^{3+5} = 2^8$
$2^5/2^3 = 2^2 = 4$
$2^3 \cdot 2^5 = 4^8$
$2^5/2^3 = 1^2$
Which road do we take here?
What do you THINK?

Of course you should try your hand at
a few more of these,
to make sure you can do them
easily.
Make them up yourself!

And you will find
Division equally easy.
Look at the following illustration.
Suppose you wish to divide:

$$a^5 \div a^3 \quad \text{or} \quad \frac{a^5}{a^3}.$$

This means

$$\frac{aaaaa}{aaa}$$

does it not?
And since these are all numbers,
you may "cancel" three of the a's, thus,

$$\frac{\cancel{a}\cancel{a}\cancel{a}aa}{\cancel{a}\cancel{a}\cancel{a}}$$

just as you would in Arithmetic;
that gives $a \cdot a$ or a^2 for the answer,

so that $\dfrac{a^5}{a^3} = a^2$.

And you no doubt can see
the justice of the following rule:
in DIVIDING a base with an exponent by
the SAME base with another exponent,
RETAIN the BASE and

SUBTRACT the EXPONENT in the divisor
from the EXPONENT in the dividend.

Now try these:

(1) $\dfrac{x^7}{x^3}$ (2) $\dfrac{a^5}{a}$ (3) $\dfrac{m^4}{m^3}$ (4) $\dfrac{b^r}{b^3}$

Did you get the following answers?

(1) x^4 (2) a^4 (3) m (4) b^{r-3}

If not,
find out why not,
so that you will be sure to
practise correctly,
when you have finished reading
this chapter.

Before starting this practice
see if you understand these examples:

$$(-5a^3b^2)(2a^2b^4) = -10^5b^6$$
$$(-3axy)(-4a^2) = 12a^3xy$$

In each case, notice that you must:
(1) Apply the rule for signs (see page 71).
(2) Multiply the NUMERICAL COEFFICIENTS
(see page 37).
And for EACH BASE SEPARATELY
(3) RETAIN the BASE (see page 81) and
(4) ADD the EXPONENTS (see page 81).

If you do each of these four steps
CAREFULLY and SLOWLY at first,
you will soon learn to do them
CAREFULLY and QUICKLY.

And, similarly for these two DIVISION examples:

$$15a^2b^4 \div (-3ab^2) = -5ab^2$$
$$-6a^m \div 2a^t = -3a^{m-t}$$

Here again you have four separate steps:
(1) Apply the rule for signs (see page 73).
(2) Divide the numerical coefficients.
And, for each base SEPARATELY
(3) RETAIN the BASE (see page 84), and
(4) SUBTRACT the EXPONENT in the divisor
from the EXPONENT in the dividend
(see page 85).

In practising
do NOT FORGET that
Times signs are sometimes UNDERSTOOD:
(1) Between a number and a letter,
thus $3a$ means $3 \times a$;
(2) Between two letters,
thus ab means $a \times b$;
(3) Between two parentheses,
thus $(-2)(-3)$ means $(-2) \times (-3)$.

And REMEMBER that

a means $1a$

xy means $1xy$

$(3ab)$ means $1(3ab)$

$-a$ means $-1a$

$-(5ab)$ means $-1(5ab)$

$(x + y)$ means $1(x + y)$.

And since in $-1(5ab)$

a times sign is understood between

the -1 and the $5ab$,

hence $-1(5ab) = -5ab$.

Similarly $(x + y) = 1(x + y) = x + y$

and $-(x + y) = -1(x + y) = -x - y$.

Do you understand these details

thoroughly?

If so, see if you can

rewrite the following,

omitting all

unnecessary signs and parentheses:

(a) $+(-6)$

(b) $+(+6)$

(c) $-(+7)$

(d) $-(-15)$

(e) (-17.3)

(f) $+(37)$

(g) $(-9) - (-8)$

(h) $(7.3) - (+6)$

(i) $-(\frac{1}{3})(\frac{2}{3})$*

---

*It is NOT advisable to write a fraction
with an oblique line, thus, 7/8, because

(j) $(-10)(-4)(\frac{7}{8})(-2)$

Did you rewrite the examples
as follows?

(a) $+(-6) = +1(-6) = -6$
(b) $+(+6) = +1(+6) = 6$
(c) $-(+7) = -1(+7) = -7$
(d) $-(-15) = -1(-15) = 15$
(e) $(-17.3) = 1(-17.3) = -17.3$
(f) $+(37) = +1(37) = 37$
(g) $(-9) - (-8) = -9 + 8 = -1$
(h) $(7.3) - (+6) = 7.3 - 6 = 1.3$
(i) $-(\frac{1}{3})(\frac{2}{3}) = -\frac{2}{9}$
(j) $(-10)(-4)(\frac{7}{8})(-2) = (40)(\frac{7}{8})(-2) = 35(-2) = -70$

Did you get the right answers?
If so, you must now practise
a sufficient number of exercises
so that you may
CORRECTLY and QUICKLY
perform the
FOUR FUNDAMENTAL OPERATIONS
with monomials containing
LETTERS and

---

when you wish to write $\dfrac{2x}{3}$ and write 2/3x,

you see that the x is
incorrectly put with the 3
instead of with the 2,
as you intended to do.

88

RATIONAL NUMBERS,
Positive and negative, and zero.*

If you want to play any game,
you must of course be familiar with
the EQUIPMENT and the RULES.
BUT
if you wish to play the game WELL,
you must practise too, you know!
That is the only way to
really ENJOY playing a game.

* Turn to pages 196 #10 and #11.

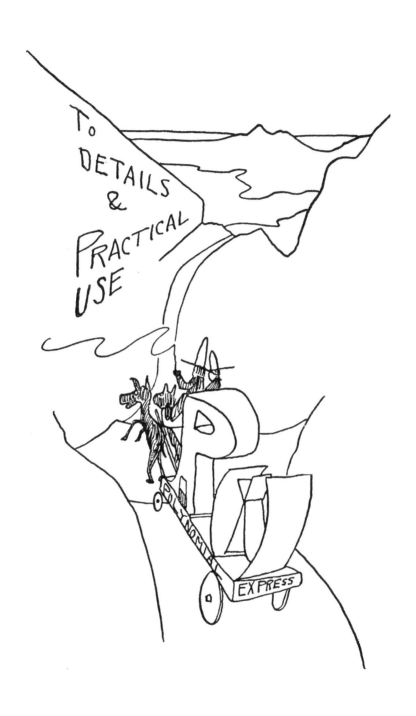

XII. WHERE DO WE GO FROM HERE?

Since we started by telling you (page 14).
that there are only TWO new ideas,
namely,
NEGATIVE NUMBERS and LETTERS,
in Algebra;
and since we told you (page 30)
that all you have to learn in Algebra is
how to perform the
FOUR FUNDAMENTAL OPERATIONS with
negative numbers and letters,
perhaps you may now ask:
"Well, we have done all this;
what else is there to it?
Why isn't this the end of the book?"

The answer is:

(1) Just as in Arithmetic,
after you have learned
to do examples like
$3 \times 7 = 21$ and $6 \div 3 = 2$,
you still have to learn how

to multiply numbers like

> 6897 by 5

and

> 83259 by 684;

and how to divide

> 7392 by 3

and

> 695 by 23

(that is, "long division");
so also, in Algebra,
you still have to learn
how to multiply and divide
algebraic expressions containing
MORE THAN ONE TERM.

And

(2) besides,
you will of course want to see
what practical use you can make
of all this Algebra which
you have learned.

XIII. ZERO

But before we go any further,
we must tell you a few things about
ZERO.

Perhaps some of you do not say "zero,"
but say "nothing" instead;
we shall soon convince you that
ZERO is really something!

For instance,
which would you rather have:
$1 or $1,000,000?
Are those zeros worth anything?
What do YOU think?

Now there are several facts about zero
which we wish to call to your
ATTENTION!

(1) If you wish to divide
    $0 among five people,
    what would each one get?

Obviously each one would get o,
would he not?
Or, if there were seven people,
or any number of people,
each one would still get o.
That is, o ÷ a = o.

(2) But suppose you wish to divide

a by o.

What is the answer now?
Do not answer hastily!
Read carefully the following
argument.
You understand all the answers
given below,
do you not?

12 ÷ 12 = 1
12 ÷ 6 = 2
12 ÷ 4 = 3
12 ÷ 3 = 4
12 ÷ 2 = 6
12 ÷ 1 = 12
12 ÷ .5 = 12 ÷ $\frac{1}{2}$ = 12 × 2 = 24
12 ÷ .1 = 12 ÷ $\frac{1}{10}$ = 12 × 10 = 120
12 ÷ .01 = 12 ÷ $\frac{1}{100}$ = 12 × 100 = 1200
12 ÷ .001 = 12 ÷ $\frac{1}{1000}$ = 12 × 1000 = 12000
and so on.

Have you noticed that

in all these little division examples
the DIVIDEND has remained the SAME,
and the DIVISOR has DECREASED from
12 to .001;
and, as a result,
the ANSWER has INCREASED from
1 to 12,000.
Now,
if we should
DECREASE the DIVISOR still further,
the ANSWER would of course INCREASE
more and more,
until, finally,
when the DIVISOR has become ZERO,
the ANSWER would become so ENORMOUS
that there is
NO NUMBER BIG ENOUGH
to express it;
we call such an answer
INFINITELY large, or
INFINITY,
and represent it by the symbol, ∞.
But remember that
∞ is NOT a NUMBER.
It is TOO BIG and TOO INDEFINITE
to be a number,
and therefore,
it is not one of the elements (see page 26),
or things we "play with" in Algebra.
That is to say,

12 ÷ 0, which is ∞,
does NOT belong to this game of Algebra.
Of course if we had chosen
some number other than 12 for
the dividend,
for example 10,
it could be shown,
by exactly the same argument,
that 10 ÷ 0 is also ∞, and therefore
also does not belong to Algebra.
And similarly,
the same would be tru of
any other number, *a*,
so that *a* ÷ 0,
being TOO BIG and TOO INDEFINITE
to be a number,
is therefore RULED OUT of Algebra.
In other words,
DIVISION BY ZERO IS RULED OUT!
(*a* ÷ 0 is a foul!)

But of course
do not confuse *a* ÷ 0, which is ∞,
with 0 ÷ *a*, which is 0 (see page 98).
Remember that 0 is a
PERFECTLY GOOD NUMBER (see page 97).
And therefore 0 ÷ *a* IS ALLOWED,
whereas *a* ÷ 0 is NOT allowed!

Perhaps some clever student will now ask,

"How about 0 ÷ 0?
Is this allowed or not?"
The answer will again be quite a surprise.

You know of course that

$$3 \times 0 = 0,$$
$$5 \times 0 = 0,$$

and, in general,

$$a \times 0 = 0,$$

where a is any number.
Now, you also know that
if a product of two numbers is
divided by either one of the numbers,
the result must be the other number:
thus,
since $3 \times 4 = 12,$
then $12 \div 4 = 3.$
Similarly, it would seem that
since $3 \times 0 = 0,$
then $0 \div 0 = 3;$
and since $5 \times 0 = 0,$
then $0 \div 0 = 5;$
and, in general,
since $a \times 0 = 0,$
then $0 \div 0 = a;$
which means that
$0 \div 0$ might equal ANY NUMBER:
3 or 5 or any other number!
And so you see that

the trouble with 0 ÷ 0 is that
it would have so MANY possible answers
that it would be very
unusual and confusing,
since, as a rule,
a division example (like 12 ÷ 3)
has ONE and ONLY ONE answer.

Later on,
When you study a branch of
Mathematics known as the
Calculus,
you will see what
wonderful use can be made of
this confusing 0 ÷ 0.
But for the present,
you do not need to worry about it.
All you have to REMEMBER NOW is that
0 ÷ a = 0, and a ÷ 0 is RULED OUT!*

By the way,
do not get all this mixed up with

$$a \div a;$$

this of course simply equals 1,
since,
if you divide any number into itself,
it goes exactly once,
does it not?

* Where a is any number other than zero.

Thus $3 \div 3 = 1$, $\quad 7 \div 7 = 1$, $\quad 100 \div 100 = 1$,
etc.
But this has nothing to do with
our discussion about zero,
except that often some people think
that $3 \div 3 = 0$,
but of course you know that this is simply
WRONG!

So you see that
ZERO is a
VERY IMPORTANT FELLOW!

XIV. THE FINISHING TOUCHES

As we said on page 94
you must now see how to handle
algebraic expressions containing
MORE THAN ONE TERM.

In the first place,
do you remember what
a "term" is?
(See page 51.)
To make quite sure that you understand
the importance of this idea,
try the following example:
What is the value of

$$7 \times 4 - 2 \div 13?$$

Perhaps you did it this way:

$$7 \times 4 = 28; \quad 28 - 2 = 26; \quad 26 \div 13 = 2;$$

and you think that 2 is the right answer.
Well, we are sorry to disappoint you
but that is WRONG!

For the understanding is that
when you evaluate an expression
you must:
(1) Find the value of EACH TERM;
(2) COMBINE SIMILAR TERMS.
Now,
the first term here is

$$7 \times 4$$

which equals 28;
and the second term is

$$-2 \div 13$$

which equals $-\frac{2}{13}$;
and combining these two results gives

$$28 - \frac{2}{13} = 27\frac{11}{13} \text{ Ans.}$$

Some people state the procedure
this way:
Whenever you have to perform
addition, subtraction,
multiplication, and division
in evaluating an expression,
you must do the
MULTIPLICATION and DIVISION FIRST,
and THEN the
ADDITION and SUBTRACTION.
And to help you remember
the ORDER of these operations,
they write MDAS and
tell you to follow
My Dear Aunt Susie.

But although this would give you
the right answer to
the example on page 104,
IT DOES NOT ALWAYS WORK.
For instance,
try it on this one:

$$12 \div 2 \times 3 + 4 \times 5 - 1.$$

If you follow
My Dear Aunt Susie
you get the WRONG answer, 21.
But if you follow
the directions on (1) and (2) page 105,
evaluating WITHIN each term from
left to right,
you get the right answer, thus:

$$18 + 20 - 1 = 37.$$

Now try these:
(1) $12 \times 5 \div 3 + 8 \div 2 - 5 \times 3 = ?$
(2) If $x = 4$, find the value of

$$2x^2 - 5x - 7.$$

Did you get 9 for the answer to (1),
and 5 for the answer to (2)?
Then you are right.
But of course you should practise
several examples of this kind

before being quite sure that
you can really do them.*

And now let us see
how to multiply and divide
expressions having more than one term.
This is similar to
multiplying numbers having
more than one digit, thus:

$$
\begin{array}{r}
78 \\
45 \\
\hline
390 \\
312 \\
\hline
3510
\end{array}
$$

As you know perfectly well,
This process consists of three steps:

(1) Multiply each digit of the multiplicand
   by 5,
   obtaining the first "partial product," 390;
(2) Similarly,
   obtain the second partial product, 312,
   by using the multiplier, 4;
(3) Add the partial products together,
   being careful to have
   "tens under tens,"
   "hundreds under hundreds," etc.,
   since you can add only "similar" things.

* Pages 198 and 199 #12 and #13.

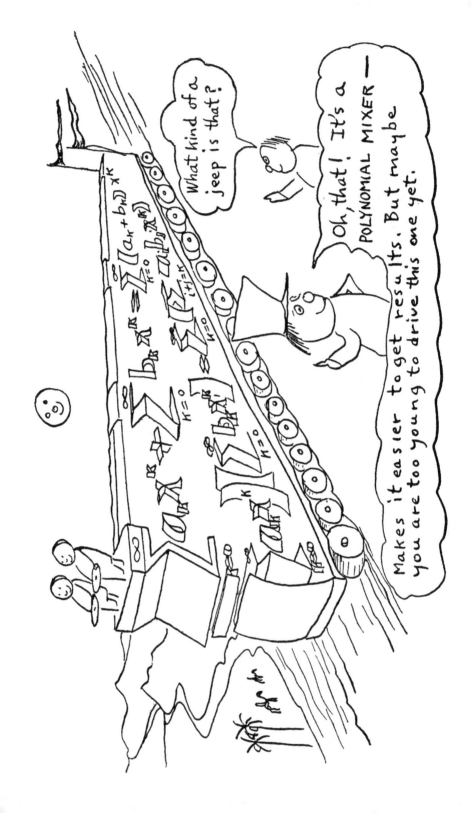

In the same way,
you can multiply two binomials, thus:

$$a + b$$
$$\underline{c + d}$$
$$ac + bc$$
$$\underline{ ad + bd}$$
$$ac + bc + ad + bd$$

in which you obtain:

(1) the first partial product, $ac + bc$;
(2) the second partial product, $ad + bd$;
(3) the sum of the partial products,
which is the final result

The only difference between this example
and the previous one, with numbers,
is that
in Algebra we work from
LEFT to RIGHT,
since we have nothing to "carry,"
as we have in Arithmetic.
Consider also the next two examples:

(1) Multiply $2x + 3$ by $3x - 1$.

$$2x + 3$$
$$3x - 1$$

$6x^2 + 9x$	first partial product
$\underline{ - 2x - 3}$	second partial product
$6x^2 + 7x - 3$	the sum of the partial products

Note here that
when adding the partial products,
you arrange similar terms* in
the same column,
just as you do in Arithmetic (page 107),
and therefore
the answer here
contains only three terms instead of four.

(2) Multiply $4x - 5y$ by $4x + 5y$.

$$
\begin{array}{r}
4x - 5y \\
\underline{4x + 5y} \\
16x^2 - 20xy \\
\underline{ 20xy - 25y^2} \\
16x^2 - 25y^2
\end{array}
$$

Note here that
the answer contains only two terms,
since the middle one dropped out.

---

* On page 52, we promised you
  a better definition of "similar terms."
  For terms to be "similar,"
  it is not enough for them to have
  the same letters,
  but
  the EXPONENTS on each letter
  must also be the same:
  For instance,
  $5x^3y^2$ is similar to $7x^3y^2$
  but not to $7x^2y$;
  also,
  $6x^2$ is NOT similar to $9x$
  and must therefore be put
  in a separate column.

Obviously this will always happen when
the two binomials are exactly alike EXCEPT
for the sign between the terms,
one being plus, the other minus;
such binomials are known as
"conjugates."
Thus $x + y$ and $x - y$ are conjugates,
and their product is merely $x^2 - y^2$,
also, $3x + 2y$ and $3x - 2y$ are conjugates,
and their product is simply $9x^2 - 4y^2$.

Thus, you see that
the product of two binomials
may have four, three, or two terms,
as shown above.
And now it is easy to see that
even if the multiplicand and the multiplier
have MORE than two terms,
the above method of
obtaining the partial products and
adding them
will apply here also.
Thus, try your hand at the following one:
Multiply $5x^3 - 2x^2y + 3xy^2 - 7y^3$
by $x^2 + 4xy - 6y^2$.*

Perhaps the thoughtful student will ask:
"But why do you ADD the partial products,

* You are not going to look up the answer on page 220
  BEFORE you have done the example yourself,
  ARE you?

when it is really a
MULTIPLICATION example?"

To which we reply that
it is very easy to understand this
if you remember the basic RULES (page 24).
Thus,
applying the Distributive Law to

$$(a + b)(c + d)$$

we get

$$(a + b)c + (a + b)d;$$

then,
applying the Commutative Law for Multiplication
and the Distributive Law again,
in each term,
we finally get

$$ac + bc + ad + bd,$$

the same answer we got on page 109.

Similarly,
applying this Law to $(2x + 3)(3x - 1)$,
we get

$$(2x + 3)3x + (2x + 3)(-1),$$

and then

$$6x^2 + 9x - 2x - 3;$$

And combining similar terms,
this becomes

$$6x^2 + 7x - 3,$$

as before.
So you see
that we are merely using
the basic RULES
even in these more difficult cases!
Since $(2x + 3)(3x - 1)$ has been shown
to be equal to $6x^2 + 7x - 3$,
we can say that
$(2x + 3)$ and $(3x - 1)$ are
the FACTORS of $6x^2 + 7x - 3$.
(See page 25.)
Similarly,
$(x + y)$ and $(x - y)$ are
the FACTORS of $x^2 - y^2$, and
$(3x + 2y)$ and $(3x - 2y)$ are
the FACTORS of $9x^2 - 4y^2$.
You will of course realize
that it is easier to do
a multiplication example
(as shown on page 110)
than to go in reverse
and FIND the FACTORS
when the PRODUCT is given.
For instance,
can you see that
the FACTORS of $x^2 - 4$
are $(x + 2)$ and $(x - 2)$?
Can you prove it?
Can you factor $100a^2 - b^2$?
And how about $x^2 - 5x + 6$?

Do not worry if
you cannot factor these yet.
We are coming back to this later!
But for the present,
you had better practise
some multiplication
(pages 199 and 200, #14 and #15).
Finally
you will see also that
LONG DIVISION in Algebra is
very similar to
LONG DIVISION in Arithmetic.
Thus, consider first
The various steps in the following
Arithmetic problem:
Divide 69872 by 324.

$$215\tfrac{212}{324} = 215\tfrac{53}{81} \text{ Ans.}$$

$$
\begin{array}{r}
324)\overline{69872} \\
\underline{648} \\
507 \\
\underline{324} \\
1832 \\
\underline{1620} \\
212
\end{array}
$$

Step (1) Divide 3 into 6 to get
the "trial" quotient, 2.

Step (2) Multiply the ENTIRE divisor, 324,
by the quotient, 2,
obtaining 648.

Step (3) Subtract 648 from 698,
obtaining 50.

Step (4) Bring down the next figure, 7.

Then REPEAT these four steps
until there are no more figures
to bring down.
Here we have a remainder, 212,
and the answer is $215\frac{53}{81}$.

Now compare this,
step for step,
with the following
long division example in Algebra:
Divide $x^2 + 5x - 6$ by $x + 2$.

$$
\begin{array}{r}
x + 3 \qquad\quad \text{quotient} \\
x + 2)\overline{x^2 + 5x - 6} \\
\underline{x^2 + 2x} \qquad\qquad\; \\
3x - 6 \\
\underline{3x + 6} \qquad\quad\;\; \\
-12 \quad \text{remainder}
\end{array}
$$

Step (1) Divide x into x^2 to get
the first term of the quotient, x;

here, however,
it is not a "trial" divisor,
since it will never be too large,
as sometimes happens in Arithmetic,*

* Thus in the example on page 114
when dividing 3 into 18,
the trial divisor 6 is too large,
and we take 5 instead.

making the process a little
easier in Algebra than in Arithmetic.

Step (2) Multiply the ENTIRE divisor $x + 2$ by x,
obtaining $x^2 + 2x$.

Step (3) Subtract $x^2 + 2x$ from $x^2 + 5x$,
obtaining $3x$.

Step (4) Bring down the next term, -6.

Then REPEAT these four steps.
Usually the remainder is merely labeled, as shown.

Of course both these examples
may be checked
by
(1) multiplying the divisor by the quotient, and
(2) adding in the remainder;
the result should be the dividend.
This method of checking may
conveniently be written
in Algebra language, like this:

$$D = dQ + r,$$

where D = dividend,
d = divisor,
Q = quotient,
r = remainder.

This, as you see, is
a useful formula.

One point should be emphasized
in connection with

long division in Algebra,
namely,
sometimes the terms must be
rearranged
before proceeding with the division.
Thus,
to divide
$5x + x^2 - 6$ by $x + 2$,
you must first rearrange the dividend
to read $x^2 + 5x - 6$,
then proceed as before—
that is,
arrange the terms in a
regular order with respect to
the exponents of any chosen letter,
which is NOT the case in

$$5x + x^2 - 6,$$

since here the exponents of x are
in the order 1, 2, 0,*

* Note that the division example $\dfrac{a^3}{a^3}$

may be done in TWO different ways:
(1) by canceling like this,

$$\frac{a^3}{a^3} = \frac{\cancel{a} \times \cancel{a} \times \cancel{a}}{\cancel{a} \times \cancel{a} \times \cancel{a}} = 1,$$

(2) by following the procedure on page 84,

we get $\dfrac{a^3}{a^3} = a^0$.

Therefore, since 1 and a^0

each equal the SAME thing, namely $\dfrac{a^3}{a^3}$,

instead of 2, 1, 0,
as it should be.
If this is not done,
the result obtained is
equivalent but not as convenient in form.
If there is more than one letter present,
choose any one,
and proceed as above;
thus, to
divide $2ab + a^2 + b^2$ by $b + a$,
arrange either

(1) $a + b\overline{)a^2 + 2ab + b^2}$

or

(2) $b + a\overline{)b^2 + 2ab + a^2}$

Obtaining the same answer in each case.

---

they must be equal to each other, thus:
$$a^0 = 1.$$
And of course by the same reasoning,
$$x^0 = 1.$$
So, the last term of $5x + x^2 - 6$,
namely, -6,
may be considered to be $-6x^0$.
That is why we say above
that the exponents of x in the three terms
are in the order 1, 2, 0.
If you do not understand this clearly,
come back to it again some time—
you will be surprised to find that
all of a sudden, some day,
you will understand it.
Never give up if
you do not understand something
the first time you read it.
Just take it easy,
but come back to it
again and again
until you get it.

Just one more caution:
If you must divide $x^3 - y^3$ by $x - y$,
arrange the work as follows,
so as not to disturb
the order mentioned above:

$$
\begin{array}{r}
x^2 + xy + y^2 \\
x - y \overline{)x^3 \qquad\quad\; - y^3} \\
\underline{x^3 - x^2y} \\
x^2y \qquad\quad - y^3 \\
\underline{x^2y - xy^2} \\
xy^2 - y^3 \\
\underline{xy^2 - y^3}
\end{array}
$$

And, now,
After some practice in
long division,*
you may heave a sigh of relief
because
you have now acquired
the necessary technique
to appreciate the power of
the Algebra-Machine
to solve many practical problems.

* See page 201 #16.

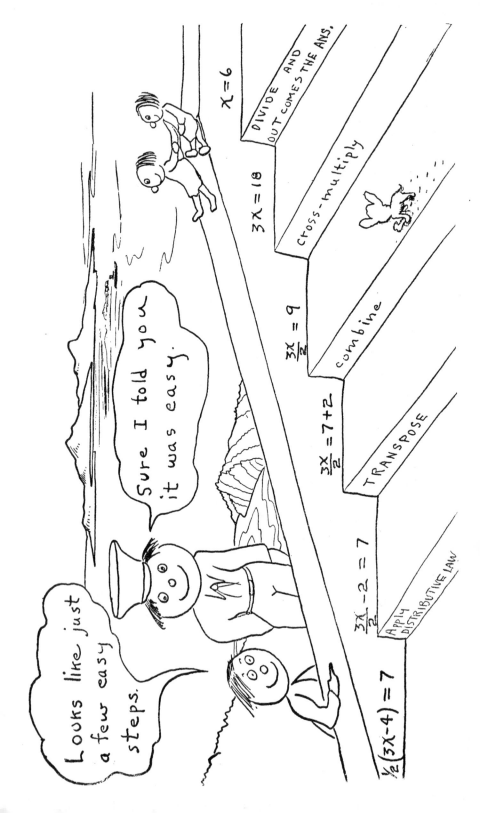

XV. A FEW EASY STEPS

On page 15
we referred you to
some useful "formulas."
And on page 116
we showed you
a formula giving the relationship between
the dividend, divisor, quotient, and remainder
of a division example.
And so you have some idea that
formulas are useful and compact.
But you probably do not realize yet
how you can actually solve
practical problems.

In order to do this,
we must tell you something about
EQUATIONS,
and, of course,
all you have done so far
will come in very handy.

In the first place,
an equation is
a statement that two things are equal.
For instance,

$$(1)\ 2 + 7 = 9$$
$$(2)\ 3x + 5x = 8x$$
$$(3)\ 7y = 21$$

are all equations.
Note, however, that
(3) is true ONLY IF $y = 3$,
whereas (2) is true NO MATTER
what value x has!
Hence (3) is called
an "equation of condition,"
since it is true only on condition
that y has a certain value,
whereas (2) is called
an "identity,"
since both sides are
identically equal for ALL values of x.
Now,
in solving practical problems
you will have to use equations;
and to solve equations,
you must bear in mind
one simple principle:
If you have two equal things,
and you wish them to stay equal,
you must treat them both alike!

Thus, if you have an amount $x + 2$
on the one hand,
and if you have 10 on the other,
and if these are equal, thus,

$$x + 2 = 10;$$

then
if you take 2 away from each,
the remainders will be equal, thus,

$$x = 8.$$

Do you understand this clearly?
And do you realize that
by doing this
you have found out the value of x?

Similarly,
if you start with the two equals,

$$x - 5 = 10,$$

you may ADD 5 to each,
obtaining equal results,

$$x - 5 + 5 = 10 + 5$$

or

$$x = 15.$$

Or,
if you start with

$$7y = 14$$

and divide each side by 7,

you get
$$y = 2.$$
And finally,
if you have
$$\frac{b}{4} = 3,$$
you may multiply each side by 4,
obtaining
$$b = 12.$$
Thus, you see that
in the four types given above,
you may obtain the value of
the "unknown" letter,
by
ADDING, SUBTRACTING,
MULTIPLYING, OR DIVIDING BY
THE SAME AMOUNT
ON BOTH SIDES OF THE EQUATION.

This is a very simple but
very useful tool for
solving equations.
Try the following examples,
being sure that you understand
when you must add, subtract, etc.
Solve:

(1) $20x = 40$

(2) $6b = 3$

(3) $\frac{y}{2} = 5$

(4) $a - 7 = 7$

(5) $c + 4 = 11$

(6) $9m = 45$

Which of the following answers are right?

(1) $x = 2$ (4) $a = 0$
(2) $b = 2$ (5) $c = 7$
(3) $y = 2\frac{1}{2}$ (6) $m = 5$

Now of course,
all equations are not so simple as these;
you may have to take
several steps to solve some of them.
For instance:

(7) Solve $5x + 3 = 18$.
First subtract 3 from both sides,
obtaining $5x = 15$;
then divide both sides by 5;
hence, $x = 3$ is the answer.

(8) Solve $\frac{y}{2} - 7 = 11$.

First add 7 to both sides, obtaining

$$\frac{y}{2} = 18;$$

then multiply both sides by 2;
hence, $y = 36$ is the answer.

(9) Solve $.2\, a = 5$.
First multiply both sides by 10,
obtaining $2a = 50$;
then divide both sides by 2;
hence, $a = 25$ is the answer.

(10) $7x + 2x = 63$.

Combining similar terms, we get

$$9x = 63;$$

then divide both sides by 9;
hence, $x = 7$.

(11) $\dfrac{5m}{2} - 3 = 7$.

First add 3 to each side, obtaining

$$\frac{5m}{2} = 10;$$

then multiply both sides by 2;
we get $5m = 20$;
and finally, divide both sides by 5;
hence, $m = 4$.

Let us now show how you can
ADD or SUBTRACT a quantity
on both sides of an equation
with great ease.
For instance,
suppose you have

$$x + 3 = 9$$

and you wish·to
subtract 3 from both sides,
you may of course write

$$x + 3 - 3 = 9 - 3$$

which becomes
$$x = 6.$$
But notice that
since the 3's on the left cancel,
we can go from
$$x + 3 = 9$$
to
$$x = 9 - 3$$
and it looks AS IF
you took the 3 on the left and
placed it on the right,
with its sign changed from + to −.
Now of course,
you did nothing of the kind!
You know that what you really did was
to SUBTRACT 3 from BOTH sides,
following the principle that
if you treat equals alike
they remain equal.
But of course
if you should carry over or "transpose"
the 3 from one side to the other,
changing its sign in the process,
you would get the same answer;
and so,
although "transposing" is only a trick,
still, since it gives the same answer,
it is permissible and
It sometimes greatly simplifies the work.
Thus,

solve $5x - 2 = 14 - 3x$
by transposing the -2 to the right
and the $-3x$ to the left;
we get

$$5x + 3x = 14 + 2;$$

combine terms, obtaining

$$8x = 16;$$

dividing both sides by 8, we get

$$x = 2.$$

BUT if you wish to see
what is the common sense behind
this mechanical procedure,
let us do it the other way, thus:
starting with $5x - 2 = 14 - 3x$,
first add 2 to each side, obtaining

$$5x = 16 - 3x;$$

next, add $3x$ to both sides, we get

$$8x = 16$$

and finally $x = 2$, as before.
In this way you see that
at each step
we merely treat both equals alike,
thus keeping them equal.
Perhaps you do not see any advantage
in the "transposing" method;
But the fact is that in cases where

you can transpose several terms at the same time,
it can be done more mechanically
and therefore more easily.
For instance,
solve $7y - 3 + 2y + 5 = 11 - 4y + 8$.
Transposing ALL terms containing y to the LEFT,
and ALL other terms to the RIGHT,
we get
$$7y + 2y + 4y = 3 - 5 + 11 + 8;$$
combining terms
$$13y = 17$$
and
$$y = \tfrac{17}{13}.$$

Try this example the other way also.

And now let us try
this more complicated equation.
Solve:
$$(m - 2) - 2(3m - 4) = -(5m + 12) + 6(m + 1).$$
First, apply the Distributive Law to each term:
$$m - 2 - 6m + 8 = -5m - 12 + 6m + 6;$$
transposing, we get
$$m - 6m + 5m - 6m = 2 - 8 - 12 + 6;$$
combining,
$$-6m = -12;$$
dividing both sides by -6,
$$m = 2.$$

All these examples can of course
be checked by
substituting the value obtained
in place of the "unknown" in
the original equation.
Thus, in the last example,
replacing every m by 2, we get

$$(2 - 2) - 2(6 - 4) = -(10 + 12) + 6(2 + 1),$$

which becomes

$$(0) - 2(2) = -(22) + 6(3)$$

or

$$0 - 4 = -22 + 18$$

or

$$-4 = -4.$$

And, since we are trying to find
a value of m which will make

$$(m - 2) - 2(3m - 4)$$

equal to

$$-(5m + 12) + 6(m + 1),$$

you see that we have succeeded,
since $m = 2$ really does make
these two quantities equal.

Now try this one.
Solve:

$$\frac{x}{2} = \frac{x + 7}{5}.$$

If we multiply both sides of this equation
by the product of the denominators, 10,
we get
$$5x = 2(x + 7).$$

Do you see this clearly?
Be sure you do before you go on.
And of course be sure that
you realize the advantage of
having thus got rid of fractions.
Now, continuing in the usual manner,
we get
$$5x = 2x + 14;$$
transposing,
$$5x - 2x = 14;$$
combining,
$$3x = 14;$$
dividing,
$$x = \tfrac{14}{3} \text{ or } 4\tfrac{2}{3} \text{ Ans.}$$

Checking, we should find that

$$\frac{4\tfrac{2}{3}}{2} = \frac{4\tfrac{2}{3} + 7}{5};$$

to simplify both sides,
we shall use the following
IMPORTANT PRINCIPLE about FRACTIONS:
If the numerator and denominator
of any fraction
are multiplied (or divided) by
the SAME number,

the value of the fraction remains unchanged.
You are undoubtedly familiar with
this principle,
but perhaps you do not use it often enough!
Applying this idea in the check above,
we multiply top and bottom on the left by 3,
obtaining $\frac{14}{6}$ or $\frac{7}{3}$;
and applying it on the right, we get

$$\frac{14 + 21}{15} \text{ or } \frac{35}{15} \text{ or } \frac{7}{3};$$

hence,

$$\frac{14}{6} = \frac{14 + 21}{15}$$

or

$$\tfrac{7}{3} = \tfrac{7}{3};$$

hence, the value $x = \frac{14}{3}$ is correct.

And now,
before we go to the practical applications,
let us call to your attention
another "trick," by means of which
you can solve equations like the last one
mechanically, but with greater ease.
You notice that you can
"clear of fractions," as above,
by merely "cross-multiplying,"
that is,
multiplying across, like this:

obtaining
$$5x = 2(x + 7).$$
Now, here again,
as in the case of "transposing,"
this is NOT what is ACTUALLY done;
for, as we said above,
what was actually done was that
EACH side was multiplied by 10, thus,

$$10\left(\frac{x}{2}\right) = 10\left(\frac{x + 7}{5}\right);$$

but since the 10 is the product of
2 and 5,
when you cancel the 2 into the 10,
you will of course get $5x$ on the left
just as if you had "cross-multiplied";
and similarly on the other side.

Thus you must see clearly that:
(1) You should UNDERSTAND
    the basic principles you use—
    as, for example,
    if you treat equals alike,
    they will stay equal.
And
(2) You should not be afraid to use
    mechanical short cuts
    (like "transposing" and "cross-multiplying")
    provided that
    you have been convinced that

133

they are really in agreement with
the basic principles.

And now,
after you have had some practice
in solving equations,*
you will then easily see
how they can be used
in solving practical problems

* On page 201 #17.

XVI. SOME PRACTICAL PROBLEMS

Let us take, for example,
the following problem:
"An aircraft flies 1 hour 30 minutes at a
certain average ground speed, and 2 hours
at 150 mph.   The entire distance traveled
was 630 miles. Find the average ground
speed for the first part of the flight."
Now this is a very simple problem,
and many people can do it by using
a little simple Arithmetic,
without any help from Algebra.
But, on the other hand,
many people can NOT do it,
and would appreciate this "tool"—
just as some people can
drive a nail with a stone,
but others find it easier to
use a hammer,
a tool which gives them the advantage of
"leverage."
It is the same with Algebra.

Now, in the first place,
many people get confused
before they give themselves a chance
to use any tool,
whether it be Arithmetic or Algebra.
And this is because they have trouble in
ANALYZING the problem—
that is,
they do not clearly see
the relationship between
the various statements in the problem,
and therefore do not know
"how to begin."

For this reason we want to show you,
first,
how to take the problem apart,
and carefully
to "pigeonhole" each little fact
in its proper place;
then you will soon begin to see
how the various facts are related,
and what to do with them.
Let us illustrate from the problem above.
Since this is a "motion" problem,
and since we know that
the product of TIME by the AVERAGE RATE
gives the DISTANCE covered,
that is,

$$d = rt,$$

let us arrange the facts as follows:

	r	$\times$	t	$=$	d
First part of flight					
Second part of flight					

We can now fill in the data, thus:

	r	$\times$	t	$=$	d
First part of flight	x		$\frac{3}{2}$		
Second part of flight	150		2		

where 1 hour 30 minutes has been changed to
$1\frac{1}{2}$ or $\frac{3}{2}$ hours,
and x is used for the original speed which is
as yet unknown.
And now, using the formula,

$$rt = d,$$

we can fill in the remaining spaces
with ease, thus:

	r	$\times$	t	$=$	d
First part of flight	x		$\dfrac{3}{2}$		$\dfrac{3x}{2}$
Second part of flight	150		2		300

Hence we see that
the distance of the first part of the flight

is $\dfrac{3x}{2}$,

and the distance of the second part
is 300.
And since the ENTIRE DISTANCE traveled
was 630 miles,
we express this by saying

$$\frac{3x}{2} + 300 = 630,$$

which is merely the same statement
translated into
ALGEBRAIC LANGUAGE!

And now,
as soon as we have succeeded in
translating our problem into
Algebraic language,
we have at our command
all the machinery of Chapter XV,
which enables us to solve the problem
almost mechanically!
Thus we first "transpose" the 300,
obtaining

$$\frac{3x}{2} = 630 - 300;$$

combining terms, we get

$$\frac{3x}{2} = 330;$$

multiplying both sides by 2,

$$3x = 660;$$

and dividing both sides by 3, gives

$$x = 220.$$

Hence we have found out that
during the first part of the flight,
the aircraft must have flown
at the average ground speed of 220 mph
for 1 hour 30 minutes,
thus covering a distance of 330 miles;
and then, by traveling 2 hours at 150 mph,
thus covering a distance of 300 miles more,
so that the total distance traveled
really was 630 miles.

Now, of course,
some of you might have done this problem
by Arithmetic,
and perhaps even more quickly.
But that is not the point!
For, later,
you will have problems which
you CANNOT do by Arithmetic,
and therefore it will pay you
to learn how to use
this Algebra Tool!
But please do not forget that
while you are learning,
you may sometimes be a little impatient
and go back to Arithmetic
with which you are more familiar—

just as,
while a person is learning to drive a car,
he may feel that,
after all,
he can WALK around the block
faster and more easily than
he can drive!
But of course this is true
ONLY in the beginning
while you are learning.
That is naturally true
while learning to use
ANY machine or tool.

And now let us summarize
the process described above:
(1) ANALYZE your problem,
     by pigeonholing the facts
     in their proper places.
(2) Translate the problem into
     Algebraic language by
     setting up an algebraic equation.
(3) Solve the equation by
     using the machinery in Chapter XV.
(4) Check to see whether your answer
     really fulfills all the requirements
     of the problem.

These four steps must be done
slowly and carefully, and deliberately

at first;
most people do not do this,
and are quickly discouraged
as soon as they look at the problem,
because they do not "flash" the answer
immediately!

The fact is that
you must NOT EXPECT to
"flash" the answer quickly,
but you must read it SEVERAL TIMES,
each time with a different purpose in mind,
thus:

(1) Read it quickly,
just to get an idea of
what the problem is about.
Thus, in the above problem,
you quickly realize that
it is a "motion" problem,
and you therefore get ready
to place the facts where they belong,
like this:*

	r	$\times$	t	$=$	d
First part of flight					
Second part of flight					

* Later on, as problems become more difficult,
this simple arrangement will no longer work,
but you will still have to ANALYZE them,
to get a clear idea of the RELATIONSHIPS between
the various given data.

143

(2) Then you read the problem again,
but this time in order to
select the separate facts and
"pigeonhole" them, thus:

	r	$\times$	t	$=$	d
First part of flight	x		$\frac{3}{2}$		
Second part of flight	150		2		

(3) Now, using the formula

$$rt = d,$$

You easily fill in
the remaining boxes:

	r	$\times$	t	$=$	d
First part of flight	x		$\dfrac{3}{2}$		$\dfrac{3x}{2}$
Second part of flight	150		2		300

(4) Now you must read the problem
AGAIN,
this time to refresh your memory
on the fact that it says that
"The entire distance traveled was 630 miles,"
which leads to the equation

$$\frac{3x}{2} + 300 = 630.$$

And now, if you have practised
the solving of equations in Chapter XV,

you will find that
solving this equation is very simple,
leading to the answer 220 mph.

(5) Finally,
you read the problem
ONCE MORE,
this time to see whether
your answer, 220 mph,
really fits the problem.

And, in more difficult problems,
you may even have to
READ THE PROBLEM several MORE times.
If you realize this,
you will not be in too great a hurry
and get yourself in a nervous state,
which is what really prevents people
from learning how to solve these problems!

And now see if you can do this one:
"A soldier walked 15 miles and then returned
immediately in a jeep which averaged 45 mph.
The entire trip required 4 hours 5 minutes.
Find the soldier's average rate of walking."

	r	$\times$	t	$=$	d
Soldier walking	x				15
Soldier in jeep	45				15

But, if $rt = d$, then

by dividing both sides by r,
this gives $t = d/r$,
which says that
to find the time of a trip, you must
DIVIDE the DISTANCE by the average RATE.
Hence:

	r	$\times$ t	$=$ d
Soldier walking	x	$\dfrac{15}{x}$	15
Soldier in jeep	45	$\tfrac{15}{45} = \tfrac{1}{3}$	15

And, since the entire trip took $4\tfrac{1}{12}$ hour,
we get the equation:

$$\frac{15}{x} + \frac{1}{3} = \frac{49}{12}.$$

Multiplying both sides by
the least common denominator, $12x$,
the equation becomes

$$180 + 4x = 49x.$$

Transposing the $4x$, we get

$$180 = 49x - 4x,$$

or $180 = 45x$;
Dividing both sides by 45,

$$4 = x,$$

or $x = 4$ mph, the answer.
Now, to check this answer:

At 4 mph
the soldier walks the 15 miles
in 3¾ hours;
but he rides back in ⅓ hour,
so that his total time is

$$3\tfrac{3}{4} + \tfrac{1}{3}$$

or

$$\tfrac{15}{4} + \tfrac{1}{3}$$

or

$$\frac{45 + 4}{12} = \frac{49}{12} = 4\,\frac{1}{12}\ \text{hours,}$$

which is the same as

4 hours and 5 minutes

for the total trip, as required.
Do you understand?

Finally, let us try one more problem.
"Eight cadets agree to buy a tent. Two of
them find that they are unable to pay, so
each one of the others has to pay $4 more
than he had expected to pay. What is the
cost of the tent?"
This, as you see, is
NOT a "motion" problem;
and yet you will be surprised to see
how similar it is.
For, here, we know that
the amount each pays, E,
multiplied by the number of people, N,
will give

the total amount of money collected, T;
hence, the formula here is:

$$EN = T,$$

which is similar to $rt = d$, is it not?
Hence we proceed as follows:

	E	$\times$ N	$=$ T
As agreed on	x	8	$8x$
What really happened	$x + 4$	6	$6(x + 4)$

But, since the price of the tent
is the same,
no matter whether
8 or 6 cadets pay for it,
and since
$8x$ and $6(x + 4)$
both represent
the COST OF THE TENT,
hence,

$$8x = 6(x + 4);$$

therefore

$$8x = 6x + 24,$$

and $8x - 6x = 24,$
or $\qquad 2x = 24;$
hence $\qquad x = 12,$
and $\qquad 8x = \$96$, the cost of the tent.
Do you see
HOW EASY IT IS?
Now try your skill on page 204 #18.

But stop for a moment
to go back to page 15 of this book
and read it again!
Do you agree now that
using letters for numbers is
really useful?
Do you realize that
being able to use x in place of
the unknown speed on page 144,
or in place of
the amount each cadet
agreed to pay (page 148),
enables us to go on and perform
all the necessary operations of
filling in the pigeonholes and of
solving the equations,
without knowing the VALUE of x
until the very end?

Do you really appreciate
what a wonderful idea this is?

XVII. GENERALIZING!

You have seen that
by using letters,
statements may be
GENERALIZED,
and thus have a
much wider use.
Thus, compare

$$a + b = b + a$$

with $\qquad 3 + 4 = 4 + 3$!

Also,
in formulas (see page 211)
you see the use of
Generalization.
Now we shall tell you
a little about Geometry
and show you how
here, too,
Generalization is very valuable.
As you know,
there are various kinds of triangles (page 55)

Now suppose you draw
any triangle you wish
(be sure to use a ruler and
draw it very NEATLY and CAREFULLY).
Then measure the three angles in it
with your protractor,
again doing this to
the best of your ability.
You will find a remarkable thing:
no matter how BIG your triangle is,
or how SMALL,
or what SHAPE it has,
the SUM of its three ANGLES
will always be
somewhere near 180 degrees!
Perhaps a little less than 180 degrees,
or a little more than 180 degrees,
but always very near 180 degrees.
Since you probably used
an ordinary protractor
(which is of course not very accurate)
and made your readings
to the nearest mark on it
(probably to the nearest degree),
the sum you find is of course
only approximate!
If you had used a better instrument,
your measurements would have been better,
and you would have found that
the sum of the three angles would have been

NEARER 180 degrees.
But of course, even with
the best instrument in the world,
measurements are NEVER perfect,
and you would *never get exactly* 180 degrees*
which it really should be!
Perhaps you will say:
"How do you know
what it SHOULD be,
if the best instruments in the world
show otherwise?"
This is certainly a fair question!
The answer is:

(1) In the first place,
     if you measure the three angles of
     the SAME triangle,
     several times with
     the SAME instrument,
     you will get
     a slightly different result each time,
     showing that the result obtained with
     the instrument is really NOT perfect!
(2) Later on,
     when you study Geometry properly,
     that is,
     when you learn
     the Geometry game,
     and know what are its
     elements and postulates

* Unless by accident.

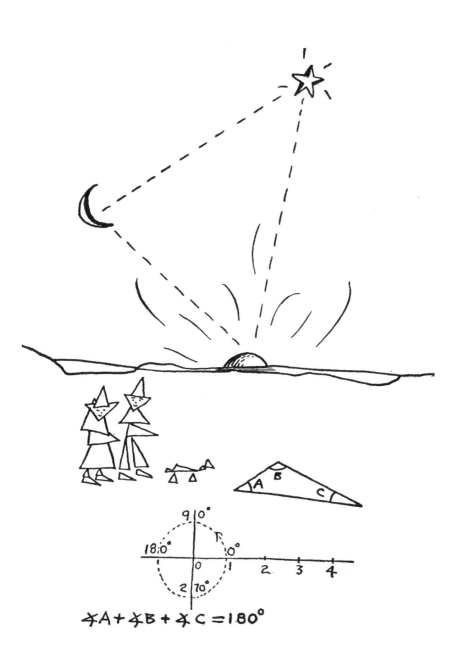

$$\angle A + \angle B + \angle C = 180°$$

(be sure to turn back to page 19!),
you will then see
how we can actually PROVE that
the sum of the three angles of a triangle
is EXACTLY 180 degrees.
But for the time being,
please take our word for it,
but do not forget to
ask the same question when
you study Geometry!

And so now the best you can do,
is to make your measurements
as carefully as possible,
repeating them several times
on the SAME triangle,
then taking an AVERAGE of
all your results—
this average being the best value
you can get at this time.

Repeat this procedure on
a very BIG triangle,
and on a SMALL one,
and compare your results
(By the way,
why is it that a triangle
CANNOT possibly have MORE than ONE
right angle in it,
nor more than one obtuse angle,

but CAN have three acute angles?)

Now draw a four-sided figure
carefully, with your ruler.
This is called a
QUADRILATERAL.
Just as with triangles,
there are VARIOUS KINDS of quadrilaterals:
Do you know the difference between
a square,
a rectangle,
a parallelogram,
a rhombus,
a trapezoid?
If not,
look up the meanings of these words
in a dictionary,*
and see if you can draw these figures.
Do you think that
the sum of the four angles of
a quadrilateral
is always the same,
regardless of its shape and size?
How much do you think this sum would be?
Let us give a hint:

* If you and your friends
  are not sure of the meanings of these words,
  please do not start a big "argument" about them,
  for that is very unscientific.
  It is so much better to
  look it up and
  really know what you are talking about.

In any quadrilateral
you can always draw a diagonal
like this:

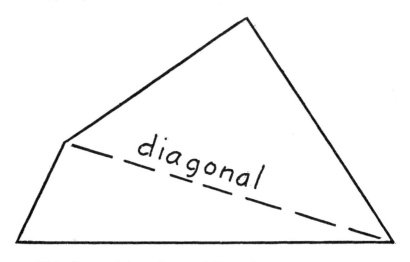

This diagonal cuts the quadrilateral
into how many triangles?
And what is the angle-sum of each triangle?
Similarly,
draw a five-sided figure
(called a PENTAGON)
and a six-sided figure
(a HEXAGON),
and, by drawing diagonals,
from a single vertex, like this:

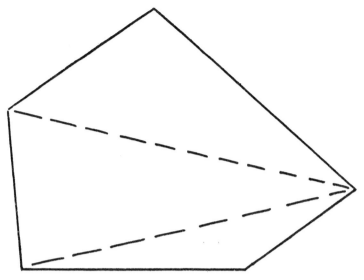

see how many triangles are formed,
and see if you can figure out
the angle-sum of these figures?
Then check with your protractor,
remembering, however, that
the protractor measurements are
only approximate.

Do you notice that
if the figure has n sides,
it is sure to contain $n - 2$ triangles?
And since the angle-sum of each triangle
is 180 degrees,
you can easily figure out
the angle-sum of

a figure having
ANY number of sides.
For instance,
a twenty-sided figure
would contain 18 triangles,
and therefore
its angle-sum would be

$$18 \times 180°$$

or 3240°.

You see that by this process of
"generalizing" from the triangle to
figures of any number of sides,
you broaden your knowledge quite easily.
The idea of generalizing is
quite popular in mathematics,
so you should be always on the watch
for possibilities of going
from something you know to
more general knowledge.
But you must be
PARTICULARLY CAREFUL
not to generalize too hastily,
for that will surely lead you astray:
for instance,
if a person has been mean to you,
do not "generalize" falsely by saying
that
all people are mean,
that it is "human nature" to be mean!

158

For, if you stop to think,
you will realize
that human nature has
some fine things in it too:
look at all the wonderful things
humans have invented,
and all the beautiful music,
etc., etc.
Human nature is
just full of good things,
if we would only
concentrate on these
instead of
on all the bad things!

XVIII. ALGEBRA AND GEOMETRY HELP EACH OTHER!

If you were asked to make
an angle of 90 degrees,
you would probably use your protractor.
Try it.
There is another way to do it,
by using a ruler and compasses,
like the sketch on page 162.

(1) First, draw the line *AB*.
(2) Put the compass-point on *A*,
and having your compasses open to
more than half the distance between *A* and *B*,
draw an arc above and below the line,
as shown.
(3) Repeat this process from point *B*
(keeping the opening of the compasses
the SAME as before),
drawing arcs crossing
the two arcs drawn in (2)
at *C* and *D*.

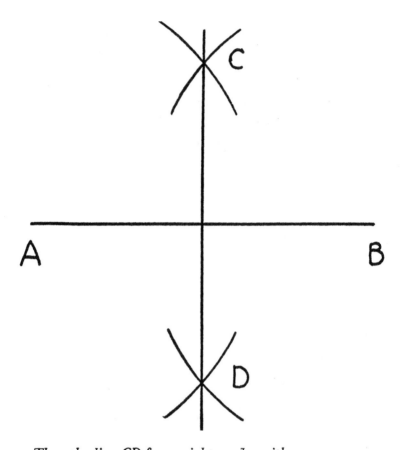

Then the line *CD* forms right angles with
the line *AB*,
and these two lines are therefore
said to be PERPENDICULAR to each other.
Not only that!
It can be shown that
CD cuts *AB* EXACTLY in half,

or "bisects" it.
Hence CD is called the
PERPENDICULAR BISECTOR of AB
All these things will be PROVED later
in your study of
Geometry.
In fact, we shall show you then that
whenever you have
two points, like C and D on page 162,
each of which is
equally distant from the ends of a line,
like AB,
(that is, $AC = BC$ and $AD = BD$),
then the line joining C and D
is sure to be
the PERPENDICULAR BISECTOR of AB.

And therefore you can modify
the construction described above
in various ways:
for instance,

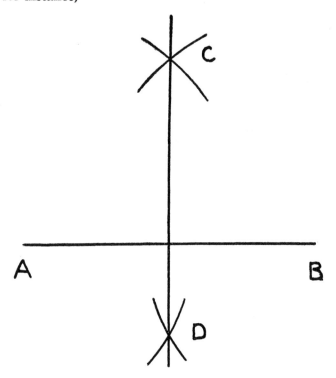

you may make the arcs as shown here,
with $AC = BC$ and $AD = BD$,
then CD is still
the perpendicular bisector of AB
(although here AC is not equal to AD,
as it was on page 162).

Or you may do it like this:

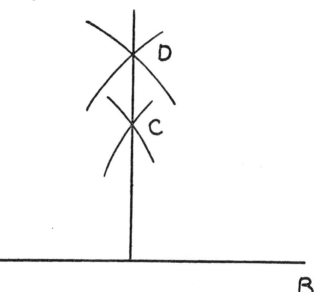

Here again we have

$$AC = BC \text{ and } AD = BD,$$

and therefore CD (if prolonged) is again the perpendicular bisector of AB.

Or

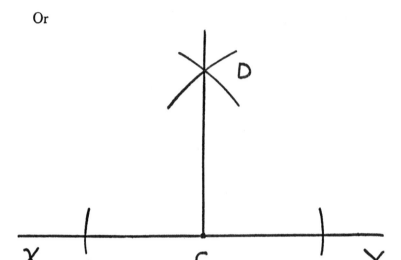

If you start with the line XY and
the point C on it,
you may again make

$$CA = CB \text{ and } AD = BD,$$

and have CD the perpendicular bisector of AB.

Or:

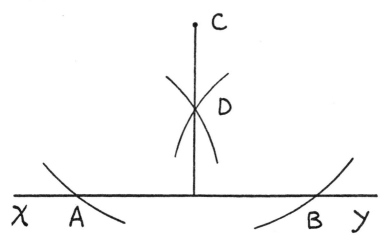

If you start with line XY and
point C outside this line,
you can still make

$$CA = CB \text{ and } AD = BD$$

so that here also
CD is the perpendicular bisector of AB.

All these ways are correct,
and sometimes one of these is
more appropriate than the others.
You should be able to do all of them.

Another construction you should
be able to do
is to cut an angle in half, that is,

to "bisect" it.
To do this,
put the point of your compasses on A,

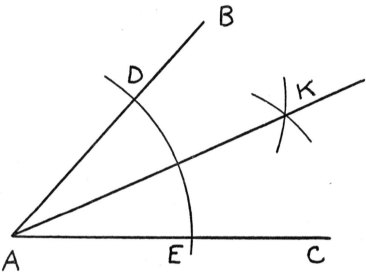

open the compasses a convenient amount,
and draw an arc cutting
the sides *AB* and *AC* as shown
at *D* and *E*.
Then from *D* and *E*
draw arcs crossing at *K* and
so that *DK* = *EK*.
Then the line *AK* will cut
the original angle in half.
(You will learn how to prove this later.)
Consequently,
if you first make a right angle (90 degrees)

by any of the methods shown above,
and if you cut it in half,
you will of course get
an angle of 45 degrees.
And, if you cut this in half again,
you can get an angle of $22\frac{1}{2}$ degrees,
and so on.
Do you see how you can now get
an angle of $67\frac{1}{2}$ degrees?*
135 degrees? 225 degrees? 315 degrees?
What other angles can you now construct
with your ruler and compasses?
Try it and
check with your protractor.

We shall also prove later that
if you have an equilateral triangle (page 56),
it is also "equiangular"
(that is, its three angles are equal to each other).
and, since the angle-sum of ANY triangle
is always 180 degrees (page 154),
each angle of an equilateral triangle
will have exactly 60 degrees.

* $67\frac{1}{2}$ is the sum of 45 and $22\frac{1}{2}$.
  Catch on?

Therefore,
if you start with a line, *AB*,
of any length:

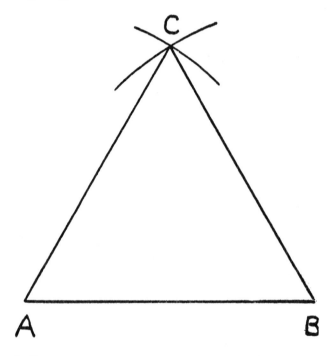

and, if you open your compasses
exactly the amount from *A* to *B*,
then draw arcs, from *A* and *B*,
intersecting at *C*,
this triangle will of course be
equilateral,
and therefore
angle *A* will have 60 degrees in it

Try it and check with your protractor.

And,
if you now bisect this angle,
you can get an angle of 30 degrees,
and from this you can get
a 15-degree angle,
and so on.

So you see,
by being able to make
a 90-degree or a 60-degree angle
to begin with,
you can from these
get a great many different kinds of angles,
using only a ruler and compasses.
How would you make
an angle of 120 degrees? 150 degrees?
What other angles can you make?
What instrument would you use to make
an angle of 59 degrees?
In the figure on page 170,
which line may be left out if
you are not interested in
drawing a complete equilateral triangle,
but want merely
one angle of 60 degrees?
You will find that you can make
many interesting things
using only a ruler and compasses.

For instance,
draw a circle with your compasses,
like this:

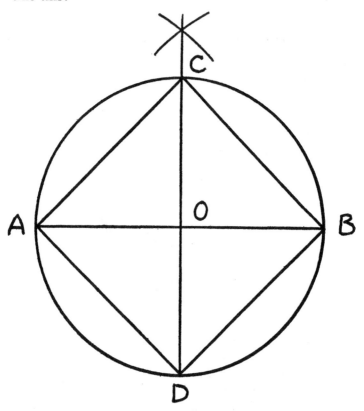

The point *0* is called its CENTER,
and the curved line is the
CIRCLE or CIRCUMFERENCE.
Now through the center draw line *AB*,
terminating in the circumference

at both ends;
such a line is called a
DIAMETER of the circle;
and half of it, like *OA* or *OB*, is called
a RADIUS of the circle.
Now construct another diameter which
is perpendicular to *AB* (page 162),
like *CD*.
Now if you draw the four lines
AC, *CB*, *BD*, *DA*,
you have a SQUARE.
And,
if you bisect the four right angles around *O*,
prolonging these bisectors until

they cut the circumference at
the four points, *E, F, G, H,*
as shown below:

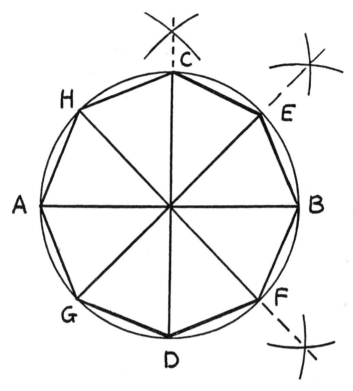

Then draw the eight lines
CE, EB, BF, FD, etc., as shown,
you will get what is known as
a REGULAR OCTAGON,
that is,
an eight-sided figure (octagon)

all of whose sides are equal,
and all of whose angles are equal
(that is why it is called REGULAR).
What other figures can you now make
with your ruler and compasses?

Let us show you how to make
a REGULAR HEXAGON (six sides):
First draw a circle:

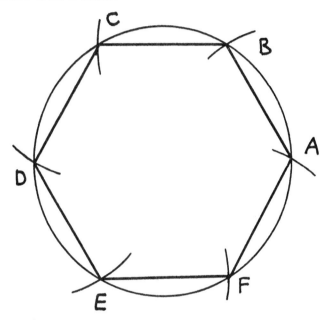

Now keep your compasses open
the SAME amount as you used
in drawing this circle,
and, putting your compass-point on A,

draw an arc cutting the circle at *B.*
Then from *B,* draw an arc at *C,*
and so on;
you will find that
the sixth time you will arrive back
at point *A.*
And, if you now join all the points
as shown,
you will have
a regular hexagon
What would you get if
you joined only the alternate points,
A, C, and *E?*
How can you now get
a regular twelve-sided figure
(a REGULAR DODECAGON)?
What other figures can you get
from your hexagon?

But perhaps you are thinking:
"What has all this to do with
Algebra?"
The answer is that
you will find later that
although Algebra and Geometry
each has its own
set of elements and
its own postulates (see page 19),
yet
they are very closely related and

many Algebraic problems
can be solved with the aid of Geometry
and vice versa.
Let us give you just a very few
simple illustrations now:

(1) What is the number of degrees in an angle which
is twice its complement? (See page 61.) Using
algebra, we say:
Let x $=$ the number of degrees in the complement.
Then $2x =$ the number of degrees in the angle itself.
And, since they are complementary
Then $x + 2x = 90$
$$3x = 90$$
and $x = 30$
hence $2x = 60$.

(2) Find the number of degrees in an angle which is
$\frac{2}{3}$ of its supplement. (See page 63.)
Let $x =$ the number of degrees in the supplement.
Then $\dfrac{2x}{3} =$ the number of degrees in the angle.
Then $x + \dfrac{2x}{3} = 180°$.
Multiplying both sides by 3, we get
$$3x + 2x = 540$$
or $5x = 540$
and $x = 108$.
Hence $\dfrac{2x}{3} = 72°$ Answer.

(3) Find the number of degrees in an angle if its complement exceeds the angle by 10 degrees.

Let x = the number of degrees in the angle.

Then $x + 10$ = the number of degrees in the comp.

Then $x + x + 10 = 90°$

or $\quad\quad x + x = 90 - 10$

or $\quad\quad\quad\quad 2x = 80$

and $\quad\quad\quad\quad x = 40°$ Answer.

(4) How many degrees are there in each of two supplementary angles which are in the ratio of 4 to 5?

Let $\quad 4x$ = the smaller angle ⎱ *

Then $5x$ = the larger angle ⎰

Then $4x + 5x = 180$

or $\quad\quad\quad 9x = 180$

and $\quad\quad\quad\ x = \ 20.$

Hence $\quad\quad 4x = \ \ 80°$ ⎱ Answer.

and $\quad\quad\ 5x = 100°$ ⎰

(5) Find the number of degrees in each of two angles which are equal and supplementary?

Let $\quad x$ = the number of degrees in one angle.

Then x = the number of degrees in the other angle.

Then $x + x = 180$

or $\quad\quad 2x = 180$

and $\quad\quad\ x = \ \ 90°$ in each.

* The ratio of these is of course $\dfrac{4x}{5x}$,

and, canceling the x,

the ratio becomes 4.5, as required.

(6) The span of the Curtiss P-40 is 37.33 feet.
The mean chord (average width of wing) is 6.32 feet.
Find the wing area.
Since the area of a rectangle is given by the formula

$$A = lw$$

where l is its length and w its width,
we may find the area of an airplane wing
by multiplying its length (span)
by its average width,
hence

$$A = 37.33 \times 6.32 \text{ square feet.}$$

Work out the answer yourself.

(7) A certain runway is $4x - 2$ feet long and
$2x - \frac{1}{2}$ feet wide. Find its area. (See page 221.)

(8) Write an algebraic formula giving the relationship
between the diameter (d) of a circle and the radius
(r). (See page 213.)

(9) Solve the formula $A = lw$ for l in terms of A and w

Be sure to turn to page 204 for
more problems to practise on.
For, as you know,
you cannot learn to do anything by
merely having it explained to you.
You must actually practise DOING it
Think of learning to drive a car,

for instance!
And it is the same with
Algebra.
You must get aboard and
learn to drive it yourself
by actual practice.

In the next book of this series
you will have
a very useful combination of
Algebra and Geometry
when you come to do
GRAPHS.
For the time being you can find
interesting graphs in
many newspapers and magazines
Why not get a scrapbook
and make a hobby of
collecting graphs?
You will be surprised
how much useful information
you will get in this way—
and so easily, too!
It is even better than stamp collecting!

XIX. TAKING STOCK

So far you have learned:

(1) Just as in any Game
you must have
(a) the equipment, and
(b) the rules
before you can start to play,
so, in any Mathematical System
you must have
(a) the elements (or "equipment"), and
(b) the postulates (or "rules").

(2) The elements which you have had
so far, in Elementary Algebra,
are
the POSITIVE AND NEGATIVE
RATIONAL NUMBERS,*
and ZERO,

* Be sure to remember that
a RATIONAL NUMBER
is one which may be expressed as
the RATIO OF TWO INTEGERS.

as shown on the following line:

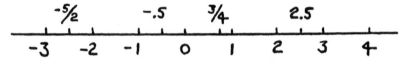

which extends to infinity in
both directions.

(3) Some of the Postulates are:
   (a) The Commutative Law for Addition:
   $$a + b = b + a.$$

   (b) The Commutative Law for Multiplication:
   $$ab = ba.$$

   (c) The Associative Law for Addition:
   $$(a + b) + c = a + (b + c).$$

   (d) The Associative Law for Multiplication:
   $$(ab)c = a(bc).$$

   (e) The Distributive Law:
   $$a(b + c) = ab + ac.$$

Etc.*

* If you are interested in
a COMPLETE set of postulates for **Algebra**,
see *The Fundamental Propositions of Algebra*
by Prof. E. V. Huntington of Harvard University,
published by
The Galois Institute Press of
Long Island University,
Brooklyn, N.Y.

184

(4) Algebraic Addition is NOT quite the same
as Addition in Arithmetic,
but has been extended to make possible
the addition of
NEGATIVE AND POSITIVE numbers:
thus

$$7 + (-11) = -4$$
$$(-7) + (-11) = -18$$

and so on.

(5) *Subtracting* a number, b,
from another number, a,
is the same as
ADDING the INVERSE of b (namely, $-b$)
to the number, a.

(6) How it follows from
the simple postulates given above,
that
in Multiplication
MINUS TIMES MINUS GIVES PLUS!

(7) The meaning of the terms:
BASE and EXPONENT,
and the following rules of
manipulating them:

$$x^m \cdot x^n = x^{m+n}$$
$$x^m \div x^n = x^{m-n}$$

(8) You must be particularly CAREFUL
in manipulating the important number
ZERO,
remembering that

$$a \times 0 = 0,$$
$$0/a = 0,$$

and $a/0$ is NOT ALLOWED.

(9) How to perform the
FOUR FUNDAMENTAL OPERATIONS
(Addition, Subtraction, Multiplication, Division)
with algebraic expressions having
ONE OR MORE TERMS.

(10) And with all this machinery,
you then saw how you could
SOLVE PRACTICAL PROBLEMS
by
expressing them in
Algebraic Language in the form of
EQUATIONS,
and solving these equations
by means of some very simple devices
such as
TRANSPOSING TERMS,
CROSS-MULTIPLYING,
etc.,
the MAIN PRINCIPLE here being that
IF YOU TREAT EQUAL THINGS ALIKE,

THEY WILL REMAIN
EQUAL TO EACH OTHER!

(11) You realize that Algebra is a
LANGUAGE
which is
PRECISE, SHORT, and
INTERNATIONAL!

(12) And you saw that,
in order to do
what is described in (10) above,
you first had to learn how to
ANALYZE a problem
and pigeonhole the various data in it
before you could set up
the necessary EQUATION.

(13) You learned how to use
some simple mechanical instruments
like compasses, protractors, rulers,
with which you can
construct and measure
various "geometric" figures.
And you will see before long
what an intimate connection there is
between Algebra and Geometry.

(14) You have learned the
importance of

GENERALIZATION (pages 20, 151, 211)
provided it is used
PROPERLY,
for it can be very
DANGEROUS if
used carelessly (page 158).

(15) You are beginning to see
how important
co-operation is,
for even Algebra and Geometry
benefit by
helping each other (page 161).

(16) And so you are
beginning to know
SAM—*
how honest he is
and intelligent
and imaginative
and helpful.
And, gradually,
whenever you run into
any difficulty,
you will learn how
to get SAM to help you
ANALYZE your difficulty
and to

* Do you remember SAM (page 4)?
There's one fellow
you must never forget!

think straight about it
instead of getting excited
and going around in circles.
Make SAM your
FRIEND and GUIDE,
and he will never desert you.
But remember that
SAM wants to be
not only YOUR friend,
but EVERYBODY'S friend,
and therefore we must
ALL GET WISE TOGETHER!

Good-by, now.
Be sure you come back for
the next episode.

$$(12x + 4y + 6z) = 2(6x + 2y + 3z)$$

XX. TRY YOUR SKILL

#1

Applying the Distributive Law:

(1) $4(5 + 2) = ?$ $20 + 8$

(2) $\frac{1}{2}(2a + 6b) = ?$

(3) $a(x + y) = ?$

(4) $2(a + b + c) = ?$

(5) $3(1 + 5x) = ?$
$3 + 15x$

(6) Factor:

$5a - 5b$ $5(a-b)$

$8x - 8y$ $8(x-y)$

$ka - kb$

$6m - 6p - 6r$ $6(m-p-r)$

$3 - 6x$ $3(1 - 2x)$

#2

Review of the FOUR FUNDAMENTAL OPERATIONS WITH POSITIVE RATIONAL NUMBERS:

(1) $17.52 + 2.895 = ?$

(2) $104.61 - 104.124 = ?$

(3) $.029 \times 17.26 = ?$

(4) $25.16 \div .258 = ?$

(5) Change $\frac{7}{15}$ to a decimal fraction, to the nearest hundredth.

(6) $\frac{9}{10} + \frac{2}{5} = ?$

(7) $\frac{7}{8} - \frac{3}{4} = ?$

(8) $9\frac{1}{2} + 2\frac{1}{3} = ?$

(9) $\frac{3}{4} \times \frac{2}{7} = ?$

(10) $\frac{3}{4} \times 2 = ?$ (13) $5\frac{1}{2} \div 2\frac{3}{4} = ?$

(11) $63 \times 2\frac{2}{9} = ?$ (14) $685 \times 27 = ?$

(12) $\frac{5}{7} \div 10 = ?$ (15) $468\frac{3}{4} \times 11.25 = ?$

(16) If a pilot flies 425 miles in 4 hours 10 minutes, how far will he fly at the same rate in 9 hours 30 minutes?

#3

Add the following:

(1) $-8 + 1 - 2 + 4 = ?$

(2) $-3 + 7 + 1 + 6 - 5 - 7 = ?$

(3) $6 - 3 + 1 - 3 + 2 - 1 = ?$

(4) $5 + 3 - 6 + 5 - 10 = ?$

(5) $2\frac{1}{2} - 3 - 4\frac{3}{4} = ?$ $-5\frac{1}{4}$

#4

Add:

(1)	(2)	(3)
$17D$	$-5q$	$+3p$
$2D$	$+3q$	$-p$

(4) $-7x + 8y - 10z + z + 8y - 3x = ?$

(5)
$6p$
$-3p$
$-p$

(6) $-5a - 4b - a - 2b = ?$

(7)
$2xyz$
$-xyz$
$-2xyz$

192

Subtract the lower quantity from the upper:

(1)	(2)	(3)	(4)	(5)
$+11n$	$+11n$	$-11n$	$-11n$	$11n$
$+\ 7n$	$-\ 7n$	$+\ 7n$	$-\ 7n$	$7n$

(6)	(7)	(8)	(9)	(10)
ab	0	ab	0	$-ab$
$-4ab$	ab	0	$-ab$	0

(11)	(12)	(13)	(14)	(15)	(16)
26	17	$7n$	$5n$	$-4x$	$-3y$
17	26	$11n$	$-11n$	$+11x$	$-3y$

#6

Add:

(1) $\quad 6a - 3b + c$
$\quad\ -3a + 2b - c$
$\quad\ \ -a + b$
$\overline{}$

(2) $4ab + 7c,\ -2ab + c,$ and $3ab - 8c.$

(3) $2x + 3y - 6,\ -2x + 8,$ and $6.$

(4) $7m - 2n + 5$ and $3 + 2n - m.$

(5) $b + g + 1,\ b - g$ and $g - 4.$

#7

(1) Subtract $2x - 3y - 4z$ from $-x + y + z.$

(2) From $3a + 2ab + 4b$ take $ab - b + 5a.$

(3) Find the difference between $a + b$ and $a - b.$

193

(4) Together a book and a pencil cost q cents. The book alone costs p cents. How many cents does the pencil cost?

(5) From the sum of $x - 9y$ and $2x + 11y$ subtract $4x - 5y + 1$.

#8

(1) If Mary is $3a$ years old now, what will represent her age b years from now? c years ago?

(2) A man's income tax is represented by $320q + 7d + 5c$. He has saved $300q + 8d - c$. How much more must he save to pay it in full?

(3) What is the perimeter of a triangle whose sides are $5y - 2x$, $6x + 3y$, and $x - 3y$?

(4) Find the perimeter of a square if one of its sides is $3a - 6b$ inches.

(5) The width of a rectangle is x feet. Its length exceeds twice the width by 5. How would you express in terms of x (a) its length, (b) its perimeter?

(6) What must be added to $2m - 5r + 6$ to make the result equal to 0?

(7) On a certain day, ten temperature readings were taken on a Centigrade thermometer. They were.

$$6°, \ -3°, \ -7°, \ -15°, \ -4°, \ 0°, \ 2°, \ 3°, \ 5°, \ 3°.$$

Find the average reading.

(8) The length of a rectangle is $x + y + \frac{1}{2}z$ feet, and its width is $x - y + \frac{1}{3}z$ feet. (a) Find its perimeter. (b) How much greater is the length than the width?

(9) Find the complement and the supplement of an angle having $2a - 3ab + 5$ degrees.

(10) If the middle one of three consecutive numbers is n, what are the other two numbers?

(11) If each of two sides of a triangle is x inches and the third side is 3 inches longer than either of the other sides, what is the perimeter in terms of x?

(12) A customer purchases 5 pounds of sugar at x cents a pound and gives in payment a dollar bill. Express in cents the change he should receive.

#9

(1) Multiply -7 by -11.

(2) $(-2)(5) = ?$

(3) $3(-7) = ?$

(4) $(-\frac{1}{2})(-\frac{2}{5}) = ?$

(5) $-\frac{1}{5}(\frac{7}{2}) = ?$

(6) $\frac{2}{3}(-\frac{5}{9}) = ?$

(7) $\frac{3}{4}(-7) = ?$

(8) $(-.15)(-2.3) = ?$

(9) $(\frac{3}{4})(8.24) = ?$

(10) (a) Add -8 and -2.

(b) From -8 take -2.

(c) Multiply -8 by -2.

(d) Divide -8 by -2.

(11) $25 \div (-5) = ?$

(12) Divide (-15) by (-2).

(13) If the area of a rectangle is $26\frac{1}{2}$, and its width is $17\frac{1}{2}$, find its length.

(14) Divide $7\frac{5}{6}$ by $2\frac{1}{3}$.

(15) $(-5\frac{1}{9}) \div (2\frac{1}{2}) = ?$
(16) $(-5)(4) \div (-2) = ?$
(17) $(-15) \div (-3) + 7 = ?$
(18) $(-22.11) \div (3) = ?$
(19) $(-1.04) \div (.02) = ?$
(20) $(5.7) \div (-19) = ?$

Find the following products:

(1) $(5a^2)(2a^7)$

(2) $(-7x^2y^3)(3xy)$

(3) $4ab^2(-a^2b)$

(4) $x^a(x^m)$

(5) $-z(z^r)$

(6) $(-3my)(-2m^2)$

(7) $c(d)$

(8) $3^2 \cdot 3^4$

(9) $(25ab)(-\frac{2}{5}a^2)$

(10) $\left(-\dfrac{3a^3b}{4}\right)^2$

(11) $\left(\dfrac{-2a^2b}{3}\right)^3$

(12) $a^2b^3 \times ab^2 \times a$

(13) $(-\frac{3}{5}xy)(\frac{1}{2}x^3y)$

(14) $4^x \cdot 4^{5x}$

(15) $2^m \cdot 2^3$

(1) Add $10a^3b^2$ and $-2a^3b^2$.
(2) From $10a^3b^2$ take $-2a^3b^2$.
(3) Multiply $10a^3b^2$ by $-2a^3b^2$.
(4) Divide $10a^3b^2$ by $-2a^3b^2$.
(5) $(-3)^3 = ?$
(6) $(-2)^2(-1)^3 = ?$
(7) $2(\frac{1}{2})^2(-4)^3 = ?$
(8) How many strips $\frac{3}{32}$ inch thick are there in a laminated piece $1\frac{7}{8}$ inches thick?

(9) Divide $\dfrac{2}{x}$ by $\dfrac{6}{x^2}$.

(10) An automobile runs at an average rate of m miles per hour. How far does it run in b hours?

(11) If one side of a square is $5x$, find (a) its area, (b) its perimeter.

(12) Frank's allowance for a week is q cents. What is his allowance for p weeks?

(13) $(-18x^5y^2) \div (-3x^3y) = ?$

(14) How many dozen oranges in a box having x oranges?

(15) Express k cents as quarters.

(16) $\dfrac{8a^5b^3}{-2ab} = ?$

(17) Divide 2 by $\dfrac{6}{k}$.

(18) Write the expression for 8 diminished by x.

(19) $(21x^3y^2) \div (-3xy^2) = ?$

(20) A boy is $3y$ years old. How old was he 5 years ago

(21) Write the formula for the total number of seats n in an auditorium if it has 2 sections, each with r rows, having s seats in each row.

(22) A basketball player scored in three games, P, S, and T points. Represent the boy's average score.

(23) Write the formula for the number of students n that may be seated in a room in which there are S single seats and T double seats.

(24) A man bought d dozen apples at c cents an apple and had 15 cents left. How many cents did he have before making the purchase?

(25) What is the width of a rectangle whose (a) area is K and length is l? (b) area is $\dfrac{7ab^2}{10cd}$ and length is $\dfrac{14b^3}{5c^2d^3}$?

(26) A group of n persons in an automobile crosses the Hudson River on a ferry. Write the formula for the total cost c if the charge is 50 cents for the car and driver, and t cents for each additional person.

(27) Write a formula for the surface area A of the 4 walls of a room whose length is L, width W, and height H.

(28) A team won x games and lost b games. What fractional part of all games played did the team win?

(29) Write a formula for the number of days n in d weeks and 3 days.

(30) $\dfrac{0}{5} = ?$ $\dfrac{7}{0} = ?$ $\dfrac{a}{a} = ?$

#12

Evaluate the following:

(1) $7 + 2 \times 11$

(2) $20 \div 2 \times 3 - 4 \div 2 \times 5 + 1$

(3) $x^2 + 7x - 11$ when $x = 2$

(4) $2x^2 - 5x + 1$ when $x = 3$

(5) $3x^2 - xy + 2y$ when $x = 2, y = 3$

(6) $\dfrac{1}{x} + \dfrac{1}{v}$ when $x = 3$ and $y = 2$

13

Using the formulas on page 211, find:

(1) The circumference and the area of a circle whose radius is 14 feet.

(2) The area of a trapezoid whose height is 7 inches and whose bases are 3 and 5 inches respectively.

(3) The volume of a box having a length of 12 inches, width $3\frac{1}{2}$ inches, and depth $\frac{3}{4}$ inch.

(4) The volume of the earth. (Its radius is about 4000 miles.) Leave the answer in terms of π.

(5) The distance an object would fall in 3 seconds.

(6) The horsepower required for an airplane if the total drag is 250 pounds and the velocity 285 mph.

(7) The Centigrade reading corresponding to:

    (a) a Fahrenheit reading of 212° (the boiling point of water).

    (b) a Fahrenheit reading of 32° (the freezing point of water).

    (c) a Fahrenheit reading of 98.6° (the normal body temperature).

(8) The time it would take for $100 to double itself at 5% (simple interest).

(9) The area of a square whose perimeter is 8x. (Hint: Find first the length of one side.)

Find the following products:

(1) $3(x^2 - 2xy)$

(2) $-2x(x + 3)$

(3) $(x + 2)(2x - 1)$

(4) $(y - 1)(y + 1)$

(5) $(3m + 5)(2m - 7)$

(6) $(a^2 + 3a + 9)(a - 3)$

(7) The length of a rectangle is $a - 2b$, and its width is $2a + b$. Find (a) its perimeter, (b) its area.

(8) One side of a square is $a - 1$. Find (a) its perimeter, (b) its area.

(9) The length of a rectangle is $a^2 + a + 1$, and its width is $a - 1$. Find (a) its perimeter, (b) its area.

(10) Does $(a - b)^2 = (b - a)^2$?

#15

In the following exercises, apply the Distributive Law to each term and then combine similar terms, as shown in the first two examples.

(1) $5(x - y) - 2(x + y) = 5x - 5y - 2x - 2y = 3x - 7y$ Answer.

(2) $(x + y) - (2x - y) = x + y - 2x + y = -x + 2y$ Answer. (See page 88.)

(3) $z(y - 3x) - z(y - x)$

(4) $x(x + y) - xy - (x^2 - 3)$

(5) $8w - (3w - 6)$

(6) $7y + 2(5y + 3)$

(7) $11x - 3(7x - 4)$

(8) $5r + (6r - 5)$

(9) $4y + 3 - (3y + 2)$

(10) $8(x - 7)$

(11) $3(z + 1) - 5(2z + 7)$
(12) $18x - (5x - 7)$
(13) $-2(a + bc) - 3(2a - bc)$
(14) $-6(8 + 2)$

#16

Perform the following divisions and check by means of the formula on page 116:

(1) $(15ab^2 - 12a^2b) \div 3ab$
(2) $(14x^3y^2 - 21x^2y^3) \div (-7x^2y)$
(3) $(a^2 + 2ab + b^2) \div (a + b)$
(4) $(c^2 + 5c + 6) \div (c + 2)$
(5) $(y^2 - 10y - 39) \div (y + 3)$
(6) $(2x^2 - 13x + 20) \div (2x - 5)$
(7) $(a^2 - 2ab + b^2) \div (a + b)$
(8) $(-xy + 2x^2 - 6y^2) \div (x - 2y)$
(9) $(x^3 - 2x^2 - 29x + 30) \div (x + 5)$
(10) $(x^3 + y^3) \div (x - y)$
(11) If the area of a rectangle is $x^3 + y^3$ and its length is $x + y$, find its width. (See page 119.)
(12) What is the remainder when $4x^2 - 4x - 2$ is divided by $2x - 1$?

#17

Solve the following equations and check by substituting the answer in the original equation as shown in the first example.

(1) Solve:

$$8x - (7x + 1) = 6$$
$$8x - 7x - 1 = 6$$
$$8x - 7x = 6 + 1$$
$$x = 7 \text{ Answer.}$$

Check:

$$56 - (49 + 1) \overset{?}{=} 6$$
$$56 - (50) \overset{?}{=} 6$$
$$6 = 6$$

(2) $5x = 15$

(3) $8y - (5y + 2) = 16$

(4) $-11z = 33$

(5) $\dfrac{25}{q} = \dfrac{5}{2}$

(6) $\dfrac{3n}{7} = 12$

(7) $5x - 3x = 22$

(8) $.07y = 1.4$

(9) $\dfrac{3z}{8} = 6$

(10) $\dfrac{m}{12} = \dfrac{15}{45}$

(11) $.7x = 1.4$

(12) $7w - 2(w - 6) = 21$

(13) $\dfrac{x}{8} = \dfrac{12}{6}$

(14) $\dfrac{2q}{7} = 18$

(15) $-13z = 52$

(16) $\dfrac{13}{q} = \dfrac{65}{5}$

(17) $.3w = .06$

(18) $11x - (7x + 4) = 40$

(19) $-21z = 42$

(20) $\dfrac{15c}{7} = 30$

(21) $10m = 5$

(22) $\dfrac{19}{k} = \dfrac{57}{15}$

(23) $7r - (6r - 5) = 7$

(24) $.35q = 7$

(25) $150x = 200(x - \frac{2}{3})$

(26) $\dfrac{3x}{2} - \dfrac{4}{3}(x - 5) - \dfrac{1}{6}(5x - 12) = 0$

(27) $4x + 5 = 2x - 7$

(28) $5x - (2x + 3) = 12$

(29) $3(9 - 2s) - 5(2s - 9) = 0$

(30) $0.25x + 17 = 0.2x - 15$

(31) $4x - (x + 0.1) = 3.8$

(32) $\dfrac{50 - 9x}{4} = \dfrac{30x - 36}{3}$

(Hint: Cross-multiply.)

(33) $12(y - 1) - 6y + 2(5 - y) = 7$

(34) $(y - 5)(y - 7) + 1 = (y - 4)^2$

(35) $8x - 1 = 24 - 2x$

(36) $\dfrac{7}{k} = \dfrac{21}{42}$

(37) $5y + 3 - (3y - 2) = 9$

(38) $.05z = 3$

(39) $-x = 20$ (What does x equal?)

(40) $8(x - 7) = -40$

#18

(1) The perimeter of a rectangle is 110 feet. Find the dimensions if the length is 5 feet less than twice the width.

$$\text{Let } x = \text{width}$$
$$2x - 5 = \text{length.}$$

Therefore $x + 2x - 5 = 55$.
(Since one width and one length go only halfway round, therefore this sum is half the perimeter.) Now solve this easy equation. Did you get the answer: width = 20 feet and length = 35 feet? Now check it.

(2) Find the dimensions of a rectangle if the perimeter is 72 inches and the width of the rectangle is $\frac{1}{2}$ the length of the rectangle. Check.

(3) Two airplanes start toward each other from two towns 680 miles apart. They meet after 2 hours. If the speed of one is 20 mph more than the speed of the other, find the speed of each.

(4) An airplane flew from town A to town B at a rate of 180 mph. It returned at a rate of 210 mph,

requiring 20 minutes less time to return. Find the distance from A to B.

(5) Mrs. Jones bought 36 yards of curtain material. She used $\frac{3}{4}$ as much in the dining room as in the living room and $\frac{1}{2}$ as much in the kitchen as in the living room. How many yards did she use in each room? Check.

(Hint: Let x = number of yards for the living room.

Then $\frac{3x}{4} + \frac{x}{2} + x = 36$.

Now multiply both sides of the equation by 4, obtaining

$$3x + 2x + 4x = 144.$$

Do you understand? And can you finish it?)

(6) A purse containing \$3.50 in quarters and dimes has 20 coins in all. Find the number of each kind of coin. Check.

(Hint: Analyze the problem as shown below:

	number of coins	value of each	total value
quarters	x	25 cents	$25x$
dimes	$20 - x$	10 cents	$10(20 - x)$

Then $25x + 10(20 - x) = 350$.

Note that all terms must be in the same denomination—here, in cents.)

(7) A purse contains 13 coins in dimes and half dollars. The total value of the coins is $3.70. Find the number of coins of each kind. Check.

(8) Two trains are 432 miles apart. If they travel toward each other at average rates of 32 and 40 miles per hour, in how many hours will they meet? Check.

(9) John buys 6 times as many Victory bonds as James, and George buys 6 more than James. In all they buy 54 bonds. Find the number purchased by each. Check.

(Hint: Let x = number of bonds that James buys.)

(10) An airplane leaves its base and flies due east at a rate of 175 mph. Fifteen minutes later another plane takes off in pursuit of the first one at a rate of 210 mph. How long does it take for the second plane to catch the first one and how far from the base does this happen? Check.

(Hint:

	r	$\times$	t	$=$	d
First plane	175		x		$175x$
Second plane	210		$x - \frac{1}{4}$		$210(x - \frac{1}{4})$

$$175x = 210(x - \tfrac{1}{4}).$$

Do you understand?)

(11) Two trains start at the same place and travel in opposite directions. In how many hours will

they be 1248 miles apart if one travels 24 mph and the other 28 mph? Check.

(12) Fred saved ⅔ as much money as John. John saved 5 times as much as Mary. Together they saved 14 dollars. How much did each save? Check.

(13) Three times a number increased by 22 is 6 more than 7 times the number. Find the number. Check. (Hint: Let x = the number. Then translate the problem into algebraic language, like this:

$$3x + 22 = 6 + 7x$$

and solve.)

(14) The length of a rectangular field is 5 more than 3 times the width. Find the dimensions if the perimeter is 82 feet. Check.

(15) One airplane can fly from San Diego to Victorville in 1½ hours. Another plane can make the trip in 2 hours. If they start from opposite ends of the course at the same time and fly toward each other, after what time will they meet?

(16) The length and width of a certain rectangle are in the ratio 3:2. Its perimeter is 30. Find its dimensions. (Hint: Let $3x$ = the length and $2x$ = the width. Why?)

(17) The perimeter of a certain rectangle is 80. If its width is 16 less than its length, find its dimensions.

(18) The width of a rectangle is ⅘ of its length. If its perimeter is 72, find its area. (Hint: Find the length and width first.)

(19) A square and a rectangle have the same area. The length of the rectangle is 8 feet more and its width is 4 feet less than a side of the square. Find the dimensions of the rectangle. (Let x = one side of the square.)

(20) The sum of 3 consecutive numbers is 93. Find the numbers. (Let x = the middle number.)

$$H = \frac{DV}{375}$$

FORMULAS

Here are some more useful formulas given in clear, brief Algebra Language and translated into more cumbersome ordinary language:

In Algebra	In English
[1] $p = a + b + c$	[1] The perimeter of a triangle is equal to the sum of its three sides.
[2] $p = 4s$	[2] The perimeter of a square is equal to four times the length of one side.
[3] $C = 2\pi r$	[3] The circumference of a circle is equal to 2π times the radius.

(π is the Greek letter corresponding to p in the English alphabet, and is pronounced "pie.")

[4] $A = lw$ or $(A = ab)$	[4] The area of a rectangle is equal to the product of its length

and width (or the product of its altitude and base).

[5] $A = \frac{1}{2}ab$

[5] The area of a triangle is equal to one-half the product of its altitude and base.

[6] $A = s^2$
(Where s^2 means ss. See page 80.)

[6] The area of a square is equal to the square of one side. (Compare this with [4] above.)

[7] $A = \frac{1}{2}b(a + b)$

[7] The area of a trapezoid is equal to one-half its height times the sum of its two bases.

(A trapezoid looks like this:

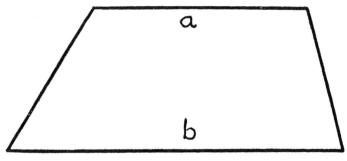

where a and b are parallel.)

[8] $A = \pi r^2$

[8] The area of a circle is equal to π times

the square of its radius.

[9] $V = lwh$

[9] The volume of a box is equal to the product of its length, width and height.

[10] $V = \frac{4}{3}\pi r^3$
(Where $r^3 = rrr$. See page 80.)

[10] The volume of a sphere is equal to $\frac{4}{3}\pi$ times the cube of its radius.

[11] $d = 2r$

[11] The diameter of a circle is equal to twice its radius.

[12] $d = rt$

[12] The distance traveled by a uniformly moving object is equal to the product of its speed and the time.

[13] $s = \frac{1}{2}gt^2$

[13] The distance covered by a freely falling object is equal to one-half the acceleration due to gravity $(g = 32)$ and the square of the time.

(Such an object does not travel "uniformly" but goes faster and faster as it falls, and that is why you cannot use Formula [12] for this.)

ANSWERS

#1

(1) 28
(2) $a + 3b$
(3) $ax + ay$
(4) $2a + 2b + 2c$
(5) $3 + 15x$

(6) $5(a - b)$
$8(x - y)$
$h(a - b)$
$6(m - p + r)$
$3(1 - 2x)$

#2

(1) 20.415
(2) 0.486
(3) 0.50054
(4) 97.5
(5) 0.47
(6) $1\frac{8}{10}$
(7) $\frac{3}{8}$
(8) $11\frac{4}{8}$

(9) $1\frac{1}{8}$
(10) $\frac{5}{8}$
(11) 140
(12) $1\frac{1}{4}$
(13) 2
(14) 18,495
(15) 5273.4375
(16) 969 miles

#3

(1) -5
(2) -1
(3) 2

(4) -3
(5) $-5\frac{1}{4}$

#4

(1) $19D$

(2) $-2q$

(3) $2p$

(4) $-10x + 16y - 9z$

(5) $2p$

(6) $-6a - 6b$

(7) $-xyz$

#5

(1) $4n$

(2) $18n$

(3) $-18n$

(4) $-4n$

(5) $4n$

(6) $5ab$

(7) $-ab$

(8) ab

(9) ab

(10) $-ab$

(11) 9

(12) -9

(13) $-4n$

(14) $16n$

(15) $-15x$

(16) 0

#6

(1) $2a$

(2) $5ab$

(3) $3y + 8$

(4) $6m + 8$

(5) $2b + g - 3$

#7

(1) $-3x + 4y + 5z$

(2) $ab + 5b - 2a$

(3) $2b$

(4) $q - p$

(5) $-x + 7y - 1$

#8

(1) $3a + b$; $3a - c$

(2) $20q - d + 6c$

(3) $5x + 5y$

(4) $12a - 24b$ inches

(5) (a) $2x + 5$

    (b) $6x + 10$

(6) $-2m + 5r - 6$

(7) $-1°$

(8) (a) $4x + 1\frac{2}{3}z$ feet

    (b) $2y + \frac{1}{8}z$ feet

(9) $-2a + 3ab + 85$;

    $-2a + 3ab + 175$

(10) $x + 1$, $x - 1$

(11) $3x + 3$ inches

(12) $100 - 5x$

#9

(1) 77
(2) -10
(3) -21
(4) $\frac{1}{8}$
(5) $-\frac{7}{10}$
(6) $-\frac{19}{27}$
(7) $-\frac{21}{4}$
(8) 0.345
(9) 6.18
(10) (a) -10
  (b) -6
  (c) 16
  (d) 4

(11) -5
(12) $7\frac{1}{2}$
(13) $1\frac{18}{35}$
(14) $\frac{47}{14}$
(15) $-\frac{92}{45}$
(16) 10
(17) 12
(18) -7.37
(19) -52
(20) $-.3$

#10

(1) $10a^9$
(2) $-21x^5y^4$
(3) $-4a^3b^8$
(4) x^{a+m}
(5) $-z^{1+r}$
(6) $6m^2y$
(7) cd
(8) 3^6

(9) $-10a^3b$
(10) $\dfrac{9a^6b^2}{16}$
(11) $\dfrac{-8a^6b^3}{27}$
(12) a^4b^5
(13) $\dfrac{-3x^4y^2}{10}$
(14) 4^{6x}
(15) 2^{m+3}

#11

(1) $8a^3b^2$

(2) $12a^3b^2$

(3) $-20a^6b^4$

(4) -5

(5) -27

(6) -4

(7) -32

(8) 20

(9) $\dfrac{x}{3}$

(10) mb

(11) (a) $25x^2$
  (b) $20x$

(12) pq

(13) $6x^2y$

(14) $\dfrac{x}{12}$

(15) $\dfrac{k}{25}$

(16) $-4a^4b^2$

(17) $\dfrac{k}{3}$

(18) $8-x$

(19) $-7x^2$

(20) $3y-5$

(21) $n=2rs$

(22) $\dfrac{P+S+T}{3}$

(23) $n=S+2T$

(24) $15+12cd$

(25) (a) $\dfrac{k}{l}$

  (b) $\dfrac{acd^2}{4b}$

(26) $c=50+t(n-1)$

(27) $A=2LH+2WH$

(28) $\dfrac{a}{a+b}$

(29) $7x+3$

(30) $0, \infty, 1$

#12

(1) 29

(2) 21

(3) 7

(4) 4

(5) 24

(6) $\frac{5}{6}$

#13

(1) $C = 88$ feet
$A = 616$ sq. ft.
(2) 28

(3) $\frac{63}{2}$
(4) $\frac{256\pi}{3} \cdot 10^9$
(5) 144 feet

(6) 190

(7) (a) $100°$
(b) $0°$
(c) $37°$
(8) 20 years
(9) $4x^2$

(1) $3x^2 - 6xy$

(2) $-2x^2 - 6x$

(3) $2x^2 + 3x - 2$

(4) $y^2 - 1$
(5) $6m^2 - 11m - 35$
(6) $a^3 - 27$

#14

(7) (a) $6a - 2b$
(b) $2a^2 - 3ab - 2b^2$
(8) (a) $4a - 4$
(b) $a^2 - 2a + 1$
(9) (a) $2a^2 + 4a$
(b) $a^3 - 1$
(10) Yes

(3) $-2xz$
(4) 3
(5) $5w + 6$
(6) $17y + 6$
(7) $-10x + 12$
(8) $11r - 5$

#15

(9) $y + 1$
(10) $8x - 56$
(11) $-7z - 32$
(12) $13x + 7$
(13) $-8a + bc$
(14) -60

(2) $x = 3$	(22) $h = 5$
(3) $y = 6$	(23) $r = 2$
(4) $z = -3$	(24) $q = 20$
(5) $q = 10$	(25) $x = \frac{8}{8}$
(6) $n = 28$	(26) $x = 13$
(7) $x = 11$	(27) $x = -6$
(8) $y = 20$	(28) $x = 5$
(9) $z = 16$	(29) $s = 4\frac{1}{4}$
(10) $m = 4$	(30) -640
(11) $x = 2$	(31) $x = 1.3$
(12) $w = \frac{8}{6}$	(32) $x = 2$
(13) $x = 16$	(33) $y = 2\frac{1}{4}$
(14) $q = 63$	(34) $y = 5$
(15) $z = -4$	(35) $x = 2\frac{5}{8}$
(16) $q = 1$	(36) $h = 14$
(17) $m = .2$	(37) $y = 2$
(18) $x = 11$	(38) $z = 6$
(19) $z = -2$	(39) $x = -20$
(20) $c = 14$	(40) $x = 2$
(21) $m = \frac{1}{4}$	

(1) $5b - 4a$	(8) $2x + 3y$
(2) $-2xy + 3y^2$	(9) $x^2 - 7x + 6$
(3) $a + b$	(10) Quotient =
(4) $c + 3$	$x^2 + xy + y^2$
(5) $y - 13$	Remainder = $2y^2$
(6) $x - 4$	(11) $x^2 - xy + y^2$
(7) Quotient = $a - 3b$	(12) -3
Remainder = $4b^2$	

#18

(2) 12; 24

(3) 160; 180

(4) 420 miles

(5) 12 yards in dining room
8 yards in kitchen
16 yards in living room

(6) 10 quarters
10 dimes

(7) 6 half dollars
7 dimes

(8) 6 hours

(9) James: 6 bonds
John: 36 bonds
George: 12 bonds

(10) $1\frac{1}{2}$ hours
$262\frac{1}{2}$ miles

(11) 24 hours

(12) Mary: $1\frac{1}{2}$
John: $7\frac{1}{2}$
Fred: 5

(13) $x = 4$

(14) 9; 32

(15) $\frac{6}{7}$ hour

(16) 9;6

(17) 28; 12

(18) 320

(19) 16; 4

(20) 30, 31, 32

Page 57:

$$\text{Perimeter} = 12xy + 9x - 10 \text{ inches}$$
$$a - b = 11xy - 9x + 1 \text{ inches}$$

Page 64:

(1) $90 - 7x + 2y$

(2) Their sum $= 90°$

(3) $m + 30 = 90$

Page 66:

(a) -63 (c) -10

(b) -8 (d) -77

Page 111:

$$5x^5 + 18x^4y - 35x^3y^2 + 15x^2y^3 - 46xy^4 + 42y^5$$

Page 118:

$$(a^2 + 2ab + b^2) \div (a + b) = a + b$$
$$(b^2 + 2ab + a^2) \div (b + a) = b + a$$

Page 179:

(6) 235.9256   (8) $d = 2r$

— (7) $8x^2 - 6x + 1$ (9) $l = \dfrac{A}{w}$

A CATALOG OF SELECTED
DOVER BOOKS
IN SCIENCE AND MATHEMATICS

Astronomy

CHARIOTS FOR APOLLO: The NASA History of Manned Lunar Spacecraft to 1969, Courtney G. Brooks, James M. Grimwood, and Loyd S. Swenson, Jr. This illustrated history by a trio of experts is the definitive reference on the Apollo spacecraft and lunar modules. It traces the vehicles' design, development, and operation in space. More than 100 photographs and illustrations. 576pp. 6 3/4 x 9 1/4.   0-486-46756-2

EXPLORING THE MOON THROUGH BINOCULARS AND SMALL TELESCOPES, Ernest H. Cherrington, Jr. Informative, profusely illustrated guide to locating and identifying craters, rills, seas, mountains, other lunar features. Newly revised and updated with special section of new photos. Over 100 photos and diagrams. 240pp. 8 1/4 x 11.   0-486-24491-1

WHERE NO MAN HAS GONE BEFORE: A History of NASA's Apollo Lunar Expeditions, William David Compton. Introduction by Paul Dickson. This official NASA history traces behind-the-scenes conflicts and cooperation between scientists and engineers. The first half concerns preparations for the Moon landings, and the second half documents the flights that followed Apollo 11. 1989 edition. 432pp. 7 x 10.
0-486-47888-2

APOLLO EXPEDITIONS TO THE MOON: The NASA History, Edited by Edgar M. Cortright. Official NASA publication marks the 40th anniversary of the first lunar landing and features essays by project participants recalling engineering and administrative challenges. Accessible, jargon-free accounts, highlighted by numerous illustrations. 336pp. 8 3/8 x 10 7/8.   0-486-47175-6

ON MARS: Exploration of the Red Planet, 1958-1978--The NASA History, Edward Clinton Ezell and Linda Neuman Ezell. NASA's official history chronicles the start of our explorations of our planetary neighbor. It recounts cooperation among government, industry, and academia, and it features dozens of photos from Viking cameras. 560pp. 6 3/4 x 9 1/4.   0-486-46757-0

ARISTARCHUS OF SAMOS: The Ancient Copernicus, Sir Thomas Heath. Heath's history of astronomy ranges from Homer and Hesiod to Aristarchus and includes quotes from numerous thinkers, compilers, and scholasticists from Thales and Anaximander through Pythagoras, Plato, Aristotle, and Heraclides. 34 figures. 448pp. 5 3/8 x 8 1/2.
0-486-43886-4

AN INTRODUCTION TO CELESTIAL MECHANICS, Forest Ray Moulton. Classic text still unsurpassed in presentation of fundamental principles. Covers rectilinear motion, central forces, problems of two and three bodies, much more. Includes over 200 problems, some with answers. 437pp. 5 3/8 x 8 1/2.   0-486-64687-4

BEYOND THE ATMOSPHERE: Early Years of Space Science, Homer E. Newell. This exciting survey is the work of a top NASA administrator who chronicles technological advances, the relationship of space science to general science, and the space program's social, political, and economic contexts. 528pp. 6 3/4 x 9 1/4.
0-486-47464-X

STAR LORE: Myths, Legends, and Facts, William Tyler Olcott. Captivating retellings of the origins and histories of ancient star groups include Pegasus, Ursa Major, Pleiades, signs of the zodiac, and other constellations. "Classic." – *Sky & Telescope.* 58 illustrations. 544pp. 5 3/8 x 8 1/2.   0-486-43581-4

A COMPLETE MANUAL OF AMATEUR ASTRONOMY: Tools and Techniques for Astronomical Observations, P. Clay Sherrod with Thomas L. Koed. Concise, highly readable book discusses the selection, set-up, and maintenance of a telescope; amateur studies of the sun; lunar topography and occultations; and more. 124 figures. 26 halftones. 37 tables. 335pp. 6 1/2 x 9 1/4.   0-486-42820-6

Browse over 9,000 books at www.doverpublications.com

Chemistry

MOLECULAR COLLISION THEORY, M. S. Child. This high-level monograph offers an analytical treatment of classical scattering by a central force, quantum scattering by a central force, elastic scattering phase shifts, and semi-classical elastic scattering. 1974 edition. 310pp. 5 3/8 x 8 1/2. 0-486-69437-2

HANDBOOK OF COMPUTATIONAL QUANTUM CHEMISTRY, David B. Cook. This comprehensive text provides upper-level undergraduates and graduate students with an accessible introduction to the implementation of quantum ideas in molecular modeling, exploring practical applications alongside theoretical explanations. 1998 edition. 832pp. 5 3/8 x 8 1/2. 0-486-44307-8

RADIOACTIVE SUBSTANCES, Marie Curie. The celebrated scientist's thesis, which directly preceded her 1903 Nobel Prize, discusses establishing atomic character of radioactivity; extraction from pitchblende of polonium and radium; isolation of pure radium chloride; more. 96pp. 5 3/8 x 8 1/2. 0-486-42550-9

CHEMICAL MAGIC, Leonard A. Ford. Classic guide provides intriguing entertainment while elucidating sound scientific principles, with more than 100 unusual stunts: cold fire, dust explosions, a nylon rope trick, a disappearing beaker, much more. 128pp. 5 3/8 x 8 1/2. 0-486-67628-5

ALCHEMY, E. J. Holmyard. Classic study by noted authority covers 2,000 years of alchemical history: religious, mystical overtones; apparatus; signs, symbols, and secret terms; advent of scientific method, much more. Illustrated. 320pp. 5 3/8 x 8 1/2. 0-486-26298-7

CHEMICAL KINETICS AND REACTION DYNAMICS, Paul L. Houston. This text teaches the principles underlying modern chemical kinetics in a clear, direct fashion, using several examples to enhance basic understanding. Solutions to selected problems. 2001 edition. 352pp. 8 3/8 x 11. 0-486-45334-0

PROBLEMS AND SOLUTIONS IN QUANTUM CHEMISTRY AND PHYSICS, Charles S. Johnson and Lee G. Pedersen. Unusually varied problems, with detailed solutions, cover of quantum mechanics, wave mechanics, angular momentum, molecular spectroscopy, scattering theory, more. 280 problems, plus 139 supplementary exercises. 430pp. 6 1/2 x 9 1/4. 0-486-65236-X

ELEMENTS OF CHEMISTRY, Antoine Lavoisier. Monumental classic by the founder of modern chemistry features first explicit statement of law of conservation of matter in chemical change, and more. Facsimile reprint of original (1790) Kerr translation. 539pp. 5 3/8 x 8 1/2. 0-486-64624-6

MAGNETISM AND TRANSITION METAL COMPLEXES, F. E. Mabbs and D. J. Machin. A detailed view of the calculation methods involved in the magnetic properties of transition metal complexes, this volume offers sufficient background for original work in the field. 1973 edition. 240pp. 5 3/8 x 8 1/2. 0-486-46284-6

GENERAL CHEMISTRY, Linus Pauling. Revised third edition of classic first-year text by Nobel laureate. Atomic and molecular structure, quantum mechanics, statistical mechanics, thermodynamics correlated with descriptive chemistry. Problems. 992pp. 5 3/8 x 8 1/2. 0-486-65622-5

ELECTROLYTE SOLUTIONS: Second Revised Edition, R. A. Robinson and R. H. Stokes. Classic text deals primarily with measurement, interpretation of conductance, chemical potential, and diffusion in electrolyte solutions. Detailed theoretical interpretations, plus extensive tables of thermodynamic and transport properties. 1970 edition. 590pp. 5 3/8 x 8 1/2. 0-486-42225-9

Browse over 9,000 books at www.doverpublications.com

Engineering

FUNDAMENTALS OF ASTRODYNAMICS, Roger R. Bate, Donald D. Mueller, and Jerry E. White. Teaching text developed by U.S. Air Force Academy develops the basic two-body and n-body equations of motion; orbit determination; classical orbital elements, coordinate transformations; differential correction; more. 1971 edition. 455pp. 5 3/8 x 8 1/2. 0-486-60061-0

INTRODUCTION TO CONTINUUM MECHANICS FOR ENGINEERS: Revised Edition, Ray M. Bowen. This self-contained text introduces classical continuum models within a modern framework. Its numerous exercises illustrate the governing principles, linearizations, and other approximations that constitute classical continuum models. 2007 edition. 320pp. 6 1/8 x 9 1/4. 0-486-47460-7

ENGINEERING MECHANICS FOR STRUCTURES, Louis L. Bucciarelli. This text explores the mechanics of solids and statics as well as the strength of materials and elasticity theory. Its many design exercises encourage creative initiative and systems thinking. 2009 edition. 320pp. 6 1/8 x 9 1/4. 0-486-46855-0

FEEDBACK CONTROL THEORY, John C. Doyle, Bruce A. Francis and Allen R. Tannenbaum. This excellent introduction to feedback control system design offers a theoretical approach that captures the essential issues and can be applied to a wide range of practical problems. 1992 edition. 224pp. 6 1/2 x 9 1/4. 0-486-46933-6

THE FORCES OF MATTER, Michael Faraday. These lectures by a famous inventor offer an easy-to-understand introduction to the interactions of the universe's physical forces. Six essays explore gravitation, cohesion, chemical affinity, heat, magnetism, and electricity. 1993 edition. 96pp. 5 3/8 x 8 1/2. 0-486-47482-8

DYNAMICS, Lawrence E. Goodman and William H. Warner. Beginning engineering text introduces calculus of vectors, particle motion, dynamics of particle systems and plane rigid bodies, technical applications in plane motions, and more. Exercises and answers in every chapter. 619pp. 5 3/8 x 8 1/2. 0-486-42006-X

ADAPTIVE FILTERING PREDICTION AND CONTROL, Graham C. Goodwin and Kwai Sang Sin. This unified survey focuses on linear discrete-time systems and explores natural extensions to nonlinear systems. It emphasizes discrete-time systems, summarizing theoretical and practical aspects of a large class of adaptive algorithms. 1984 edition. 560pp. 6 1/2 x 9 1/4. 0-486-46932-8

INDUCTANCE CALCULATIONS, Frederick W. Grover. This authoritative reference enables the design of virtually every type of inductor. It features a single simple formula for each type of inductor, together with tables containing essential numerical factors. 1946 edition. 304pp. 5 3/8 x 8 1/2. 0-486-47440-2

THERMODYNAMICS: Foundations and Applications, Elias P. Gyftopoulos and Gian Paolo Beretta. Designed by two MIT professors, this authoritative text discusses basic concepts and applications in detail, emphasizing generality, definitions, and logical consistency. More than 300 solved problems cover realistic energy systems and processes. 800pp. 6 1/8 x 9 1/4. 0-486-43932-1

THE FINITE ELEMENT METHOD: Linear Static and Dynamic Finite Element Analysis, Thomas J. R. Hughes. Text for students without in-depth mathematical training, this text includes a comprehensive presentation and analysis of algorithms of time-dependent phenomena plus beam, plate, and shell theories. Solution guide available upon request. 672pp. 6 1/2 x 9 1/4. 0-486-41181-8

Browse over 9,000 books at www.doverpublications.com

HELICOPTER THEORY, Wayne Johnson. Monumental engineering text covers vertical flight, forward flight, performance, mathematics of rotating systems, rotary wing dynamics and aerodynamics, aeroelasticity, stability and control, stall, noise, and more. 189 illustrations. 1980 edition. 1089pp. 5 5/8 x 8 1/4.   0-486-68230-7

MATHEMATICAL HANDBOOK FOR SCIENTISTS AND ENGINEERS: Definitions, Theorems, and Formulas for Reference and Review, Granino A. Korn and Theresa M. Korn. Convenient access to information from every area of mathematics: Fourier transforms, Z transforms, linear and nonlinear programming, calculus of variations, random-process theory, special functions, combinatorial analysis, game theory, much more. 1152pp. 5 3/8 x 8 1/2.   0-486-41147-8

A HEAT TRANSFER TEXTBOOK: Fourth Edition, John H. Lienhard V and John H. Lienhard IV. This introduction to heat and mass transfer for engineering students features worked examples and end-of-chapter exercises. Worked examples and end-of-chapter exercises appear throughout the book, along with well-drawn, illuminating figures. 768pp. 7 x 9 1/4.   0-486-47931-5

BASIC ELECTRICITY, U.S. Bureau of Naval Personnel. Originally a training course; best nontechnical coverage. Topics include batteries, circuits, conductors, AC and DC, inductance and capacitance, generators, motors, transformers, amplifiers, etc. Many questions with answers. 349 illustrations. 1969 edition. 448pp. 6 1/2 x 9 1/4.
0-486-20973-3

BASIC ELECTRONICS, U.S. Bureau of Naval Personnel. Clear, well-illustrated introduction to electronic equipment covers numerous essential topics: electron tubes, semiconductors, electronic power supplies, tuned circuits, amplifiers, receivers, ranging and navigation systems, computers, antennas, more. 560 illustrations. 567pp. 6 1/2 x 9 1/4.   0-486-21076-6

BASIC WING AND AIRFOIL THEORY, Alan Pope. This self-contained treatment by a pioneer in the study of wind effects covers flow functions, airfoil construction and pressure distribution, finite and monoplane wings, and many other subjects. 1951 edition. 320pp. 5 3/8 x 8 1/2.   0-486-47188-8

SYNTHETIC FUELS, Ronald F. Probstein and R. Edwin Hicks. This unified presentation examines the methods and processes for converting coal, oil, shale, tar sands, and various forms of biomass into liquid, gaseous, and clean solid fuels. 1982 edition. 512pp. 6 1/8 x 9 1/4.   0-486-44977-7

THEORY OF ELASTIC STABILITY, Stephen P. Timoshenko and James M. Gere. Written by world-renowned authorities on mechanics, this classic ranges from theoretical explanations of 2- and 3-D stress and strain to practical applications such as torsion, bending, and thermal stress. 1961 edition. 560pp. 5 3/8 x 8 1/2.   0-486-47207-8

PRINCIPLES OF DIGITAL COMMUNICATION AND CODING, Andrew J. Viterbi and Jim K. Omura. This classic by two digital communications experts is geared toward students of communications theory and to designers of channels, links, terminals, modems, or networks used to transmit and receive digital messages. 1979 edition. 576pp. 6 1/8 x 9 1/4.   0-486-46901-8

LINEAR SYSTEM THEORY: The State Space Approach, Lotfi A. Zadeh and Charles A. Desoer. Written by two pioneers in the field, this exploration of the state space approach focuses on problems of stability and control, plus connections between this approach and classical techniques. 1963 edition. 656pp. 6 1/8 x 9 1/4.
0-486-46663-9

Browse over 9,000 books at www.doverpublications.com

Mathematics-Bestsellers

HANDBOOK OF MATHEMATICAL FUNCTIONS: with Formulas, Graphs, and Mathematical Tables, Edited by Milton Abramowitz and Irene A. Stegun. A classic resource for working with special functions, standard trig, and exponential logarithmic definitions and extensions, it features 29 sets of tables, some to as high as 20 places. 1046pp. 8 x 10 1/2. 0-486-61272-4

ABSTRACT AND CONCRETE CATEGORIES: The Joy of Cats, Jiri Adamek, Horst Herrlich, and George E. Strecker. This up-to-date introductory treatment employs category theory to explore the theory of structures. Its unique approach stresses concrete categories and presents a systematic view of factorization structures. Numerous examples. 1990 edition, updated 2004. 528pp. 6 1/8 x 9 1/4. 0-486-46934-4

MATHEMATICS: Its Content, Methods and Meaning, A. D. Aleksandrov, A. N. Kolmogorov, and M. A. Lavrent'ev. Major survey offers comprehensive, coherent discussions of analytic geometry, algebra, differential equations, calculus of variations, functions of a complex variable, prime numbers, linear and non-Euclidean geometry, topology, functional analysis, more. 1963 edition. 1120pp. 5 3/8 x 8 1/2. 0-486-40916-3

INTRODUCTION TO VECTORS AND TENSORS: Second Edition--Two Volumes Bound as One, Ray M. Bowen and C.-C. Wang. Convenient single-volume compilation of two texts offers both introduction and in-depth survey. Geared toward engineering and science students rather than mathematicians, it focuses on physics and engineering applications. 1976 edition. 560pp. 6 1/2 x 9 1/4. 0-486-46914-X

AN INTRODUCTION TO ORTHOGONAL POLYNOMIALS, Theodore S. Chihara. Concise introduction covers general elementary theory, including the representation theorem and distribution functions, continued fractions and chain sequences, the recurrence formula, special functions, and some specific systems. 1978 edition. 272pp. 5 3/8 x 8 1/2. 0-486-47929-3

ADVANCED MATHEMATICS FOR ENGINEERS AND SCIENTISTS, Paul DuChateau. This primary text and supplemental reference focuses on linear algebra, calculus, and ordinary differential equations. Additional topics include partial differential equations and approximation methods. Includes solved problems. 1992 edition. 400pp. 7 1/2 x 9 1/4. 0-486-47930-7

PARTIAL DIFFERENTIAL EQUATIONS FOR SCIENTISTS AND ENGINEERS, Stanley J. Farlow. Practical text shows how to formulate and solve partial differential equations. Coverage of diffusion-type problems, hyperbolic-type problems, elliptic-type problems, numerical and approximate methods. Solution guide available upon request. 1982 edition. 414pp. 6 1/8 x 9 1/4. 0-486-67620-X

VARIATIONAL PRINCIPLES AND FREE-BOUNDARY PROBLEMS, Avner Friedman. Advanced graduate-level text examines variational methods in partial differential equations and illustrates their applications to free-boundary problems. Features detailed statements of standard theory of elliptic and parabolic operators. 1982 edition. 720pp. 6 1/8 x 9 1/4. 0-486-47853-X

LINEAR ANALYSIS AND REPRESENTATION THEORY, Steven A. Gaal. Unified treatment covers topics from the theory of operators and operator algebras on Hilbert spaces; integration and representation theory for topological groups; and the theory of Lie algebras, Lie groups, and transform groups. 1973 edition. 704pp. 6 1/8 x 9 1/4. 0-486-47851-3

Browse over 9,000 books at www.doverpublications.com

A SURVEY OF INDUSTRIAL MATHEMATICS, Charles R. MacCluer. Students learn how to solve problems they'll encounter in their professional lives with this concise single-volume treatment. It employs MATLAB and other strategies to explore typical industrial problems. 2000 edition. 384pp. 5 3/8 x 8 1/2. 0-486-47702-9

NUMBER SYSTEMS AND THE FOUNDATIONS OF ANALYSIS, Elliott Mendelson. Geared toward undergraduate and beginning graduate students, this study explores natural numbers, integers, rational numbers, real numbers, and complex numbers. Numerous exercises and appendixes supplement the text. 1973 edition. 368pp. 5 3/8 x 8 1/2. 0-486-45792-3

A FIRST LOOK AT NUMERICAL FUNCTIONAL ANALYSIS, W. W. Sawyer. Text by renowned educator shows how problems in numerical analysis lead to concepts of functional analysis. Topics include Banach and Hilbert spaces, contraction mappings, convergence, differentiation and integration, and Euclidean space. 1978 edition. 208pp. 5 3/8 x 8 1/2. 0-486-47882-3

FRACTALS, CHAOS, POWER LAWS: Minutes from an Infinite Paradise, Manfred Schroeder. A fascinating exploration of the connections between chaos theory, physics, biology, and mathematics, this book abounds in award-winning computer graphics, optical illusions, and games that clarify memorable insights into self-similarity. 1992 edition. 448pp. 6 1/8 x 9 1/4. 0-486-47204-3

SET THEORY AND THE CONTINUUM PROBLEM, Raymond M. Smullyan and Melvin Fitting. A lucid, elegant, and complete survey of set theory, this three-part treatment explores axiomatic set theory, the consistency of the continuum hypothesis, and forcing and independence results. 1996 edition. 336pp. 6 x 9. 0-486-47484-4

DYNAMICAL SYSTEMS, Shlomo Sternberg. A pioneer in the field of dynamical systems discusses one-dimensional dynamics, differential equations, random walks, iterated function systems, symbolic dynamics, and Markov chains. Supplementary materials include PowerPoint slides and MATLAB exercises. 2010 edition. 272pp. 6 1/8 x 9 1/4. 0-486-47705-3

ORDINARY DIFFERENTIAL EQUATIONS, Morris Tenenbaum and Harry Pollard. Skillfully organized introductory text examines origin of differential equations, then defines basic terms and outlines general solution of a differential equation. Explores integrating factors; dilution and accretion problems; Laplace Transforms; Newton's Interpolation Formulas, more. 818pp. 5 3/8 x 8 1/2. 0-486-64940-7

MATROID THEORY, D. J. A. Welsh. Text by a noted expert describes standard examples and investigation results, using elementary proofs to develop basic matroid properties before advancing to a more sophisticated treatment. Includes numerous exercises. 1976 edition. 448pp. 5 3/8 x 8 1/2. 0-486-47439-9

THE CONCEPT OF A RIEMANN SURFACE, Hermann Weyl. This classic on the general history of functions combines function theory and geometry, forming the basis of the modern approach to analysis, geometry, and topology. 1955 edition. 208pp. 5 3/8 x 8 1/2. 0-486-47004-0

THE LAPLACE TRANSFORM, David Vernon Widder. This volume focuses on the Laplace and Stieltjes transforms, offering a highly theoretical treatment. Topics include fundamental formulas, the moment problem, monotonic functions, and Tauberian theorems. 1941 edition. 416pp. 5 3/8 x 8 1/2. 0-486-47755-X